MEN'S COMPLETE HEALTH GUIDE

EXPERT ANSWERS TO THE QUESTIONS YOU DON'T ALWAYS ASK

NEIL BAUM, MD,
SCOTT MILLER, MD, MBA,
MINDI MILLER, PharmD,
AND DAVID MOBLEY, MD

Skyhorse Publi

T0035756

Skyhorse Publishing books may be purchased in bulk at special discounts for sales promotion, corporate gifts, fund-raising, or educational purposes. Special editions can also be created to specifications. For details, contact the Special Sales Department, Skyhorse Publishing, 307 West 36th Street, 11th Floor, New York, NY 10018 or info@skyhorsepublishing.com.

Skyhorse® and Skyhorse Publishing® are registered trademarks of Skyhorse Publishing, Inc.®, a Delaware corporation.

Visit our website at www.skyhorsepublishing.com.

10 9 8 7 6 5 4 3 2 1

Library of Congress Cataloging-in-Publication Data is available on file.

Cover design by David Ter-Avanesyan

ISBN: 978-1-5107-7403-2
Ebook ISBN: 978-1-5107-7982-2

Printed in the United States of America

Contents

Introduction

The demanding life of the modern male leaves very little time for self-care. With work, family, and social obligations, many men do not make their physical, mental, and emotional well-being a priority. With a rising rate of health risks and more difficult access to health care across the globe, navigating the system and the abundance of health-related information has become a mounting challenge. Men have very unique health needs. In response to these challenges, we authors have created a "head-to-toe" compendium of medical advice distilled down to the "need to know" for men and their loved ones.

Like most things in life, building the resilient male starts with information. Armed with the knowledge in this book, the reader can make the best decisions for a healthier and longer life. We begin the journey with the basics of the male anatomy and then quickly transition to a comprehensive guide to all the ailments that specifically affect men. The subsequent chapters then cover conditions that touch both males and females but with a twist—how these concerns affect men differently.

Recognizing the new model of health care delivery, we leave no stone unturned. In response to a continual shift in the "who," "where," "when," and "how" medical care is provided,

Men's Complete Health Guide brings it home with chapters on selecting a physician, understanding telemedicine, seeking timely medical advice, and home testing.

It is no secret: women live longer than men. Unfortunately, many men lack awareness and understanding of their health care needs. Unlike the typical female mindset, males are reluctant to seek health care in a timely fashion, let alone embrace preventative measures. We hope that *Men's Complete Health Guide* will encourage men to make the best decisions for a healthier and longer life.

Neil H. Baum, MD
Scott D. Miller, MD, MBA
Mindi Miller, PharmD, BCPS
David Mobley, MD

CHAPTER 1

Male Anatomy and Physiology

Most men know very little about what is going on "down there" and how to keep things working; when things are not quite right, they are often dismayed and frightened. Most likely, they will direct their focus on the penis, with little attention paid to the scrotum sack hanging in the background. Let's begin our journey exploring these treasured jewels and all of the related parts of the total package making up the male sexual anatomy.

The Penis

The penis has several specialized parts, each with its own function. However, all of these parts work in concert to provide erections, to facilitate release of semen during ejaculation, and to allow the passage of urine. The amazing male anatomy allows ejaculation and urination to use the same pipes and plumbing, but it is so beautifully organized that there is no mixing of urine and semen.

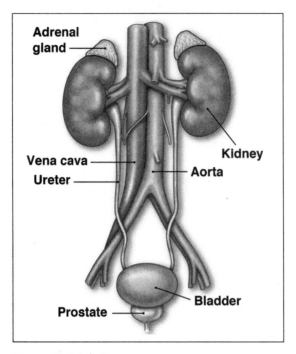

Figure 1: Male Parts

Urethra

The urethra is a very flexible, expandable tube that provides a passageway for urine to exit the bladder. The male urethra has three sections. The first section, closest to the bladder, is the prostatic urethra. In this portion, the urine passes from the bladder through the middle of the prostate gland (discussed later in this chapter.) The second section is the membranous urethra. This portion passes through the muscular pelvic floor and urinary sphincter muscle. The membranous urethra provides most of the urinary control through the activation of sphincter muscles that prevent leakage (incontinence.) The final and longest section is the pendulous, or penile, urethra.

The urethra contains small glands, called Cowper's glands, that keep the lining moist. During arousal, these glands

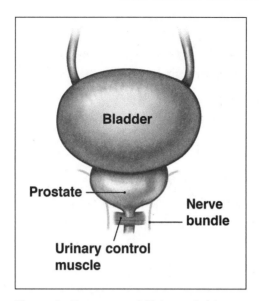

Figure 2: Prostate and Urinary Sphincter Muscle

produce additional fluid (often called "pre-cum") in order to pave the way for the ejaculation to follow. These glands can also produce additional fluid when irritated, such as with an infection. This overproduction results in a discharge from the urethra.

The final opening of the urethra is called the meatus. It is usually located at the very tip of the penis, but on occasion is on the undersurface. This uncommon condition is called hypo spadias, and it can cause a misdirected urinary stream or difficulty in achieving a pregnancy. On rare occasions, a false meatus will be located just above the normal one.

When it comes to urine flow, the urethra is much more complex than a simple hose. The portion of the urethra in the middle of the head of the penis has a slightly larger diameter than the immediately adjacent portions. This gentle disturbance in flow causes the exiting urine to spiral into a focused stream.

Narrowing of the urethra at any location can cause serious urinary difficulty. Scar tissue formation, referred to as stricture, can occur as a result of prior infection, surgery (urethra or prostate), or trauma. Meatal stenosis is a condition in which the narrowing is located at the final opening of the urethra.

Glans

The glans of the penis is most commonly referred to as the head of the penis. It is shaped like a soldier's helmet, thereby allowing smooth penetration during sexual intercourse. Its structure is very different from the remainder of the penis. The skin of the glans is unique. Although the glans expands during an erection, the skin does not stretch nearly as well as skin on other parts of the body. The glans is richly supplied with blood vessels and all types of sensory nerves (light touch, temperature, pressure, pain).

Below the skin of the glans, a specialized structure—the corpus spongiosum—fills with blood during an erection. The

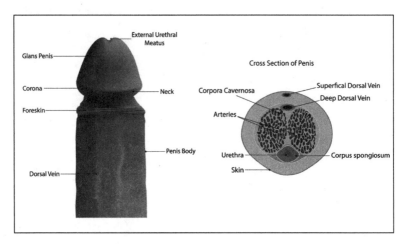

Figure 3: The Erection Mechanism

corpus spongiosum originates in the pelvis and travels as a narrow tube surrounding the urethra until it reaches the glans.

Shaft

Unlike the glans, the shaft of the penis is surrounded by very elastic skin. It is generally hairless, except for the portion closest to the body. The shaft's skin is richly supplied with nerves, but most of these are sensitive to light touch. The skin of the shaft is minimally sensitive to pain, unlike the highly sensitive glans.

The penile shaft contains two large cylinders, one on each side. These cylinders—each referred to as a corpus cavernosum—fill with blood during an erection. The outer lining of these cylinders is a tough and non-expandable layer called the tunica albuginea. Since this outer layer does not expand, once the corpus cavernosum reaches capacity, the penis becomes hard and rigid in order to allow penetration during sexual intimacy.

The corpus cavernosum is controlled by a very special nerve, aptly named the cavernosal nerve. This nerve travels along both sides of the prostate prior to exiting the body and reaching the penis. The cavernosal nerve is solely responsible for initiating an erection; it has no direct role in penile sensation or sexual climax.

Foreskin

In an uncircumcised male, the foreskin is a hood of redundant skin that covers the head or glans of the penis. During an erection, the glans usually protrudes from the foreskin. When circumcised, the redundant skin is surgically removed, leaving the glans exposed.

Scrotum

How is the scrotum like a woman's purse? It contains some very personal valuables and other things that serve a useful purpose! The scrotum contains most of the structures responsible for fertility and reproduction in the male.

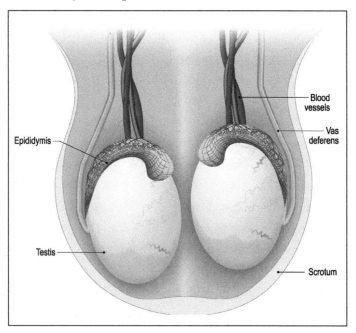

Figure 4: The Scrotum

Testicles

Although men do not pay much attention to their testicles, even a minor trauma will bring them to the forefront. Behind the scenes, a variety of specialized cells provide some essential functions. Germ cells produce sperm. Leydig cells produce testosterone. Other cells, such as Sertoli cells, serve more of a supporting role. A complex network of small tubes—the seminiferous tubules—travel throughout the testicles, allowing the sperm to reach their next destination: the epididymis.

Epididymis

Sperm travel through the epididymis during their final maturing process. The epididymis is located behind each testicle and can feel lumpy and irregular during monthly self-examination. The epididymis should also be easily distinguishable from the testicle itself. A lump in the epididymis is seldom concerning, whereas a lump on the testicle should raise the concern of testicular cancer.

Vas Deferens

The vas deferens is a very long tube that carries sperm from the epididymis to the urethra. There are two, one on each side; each takes a circuitous route from the scrotum, deep into the groin, through the pelvis, behind the bladder, and into the urethra near the prostate. It is the vas deferens that is divided through a small opening in the scrotum during a vasectomy, thereby providing an effective means of permanent birth control. Since less than 5 percent of the total volume of semen comes from the testicles, no change in semen volume is noticed during ejaculation following a vasectomy (see Chapter 6.)

Spermatic Cord

The rope-like structure that connects the testicle to the body is called the spermatic cord. It contains the blood and lymphatic vessels and nerves that supply the testicle, along with the vas deferens. The spermatic cord also contains a muscle, the cremaster muscle. This controls the elevation of the testicles in response to decrease in external temperature and other types of stimulation.

Scrotal Wall

The scrotal wall also has a muscle, the dartos muscle. It contracts in response to cold and contributes to the wrinkling effect of the scrotal skin.

Prostate

The prostate, along with two little adjacent structures called the seminal vesicles, provides only one function: manufacturing the majority of the fluid that makes up the semen. The prostate and seminal vesicles converge with the vas deferens at the urethra, thereby forming the ejaculatory duct. A mixture of all three fluids, to include the sperm from the testicles, enters the urethra at this location in order to deliver the semen in its final form to the outside world.

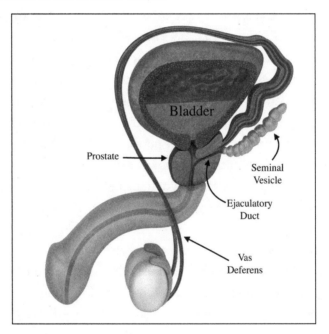

Figure 5: Ejaculatory Duct

Although the prostate does not provide any other role, it is intimately surrounded by delicate structures that supply some very important functions. These include the urinary sphincter muscle that supplies urinary control (continence) and the cavernosal nerves that initiate an erection. As a result, treatment of prostate conditions, most notably, complete removal of the prostate for cancer (radical prostatectomy), can cause damage to these adjacent structures, along with a corresponding decrease in erectile function and/or the development of incontinence (urinary leakage).

In addition, the urethra travels through the prostate as it exits from the bladder. As a result, enlargement, infection, and surgery of the prostate can alter the flow of urine. These complications will be discussed in detail.

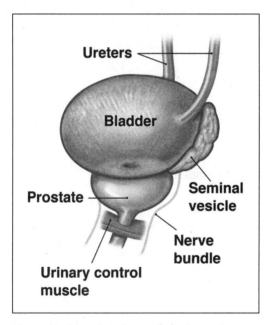

Figure 6: Hanging Around the Prostate

Bladder

The bladder serves two roles—to store and to empty the urine produced by the kidneys. We tend to take these very com plex functions for granted when all is working well. However, nerve damage or obstruction downstream in the prostate or urethra can wreak havoc on bladder function, resulting in some serious quality-of-life, and even life-threatening issues.

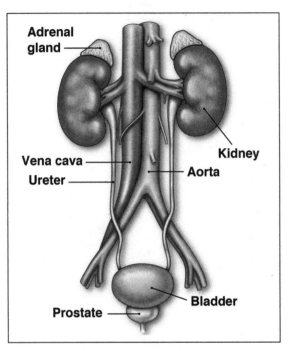

Figure 7: The Male Urinary Tract

Kidneys and Ureters

The kidneys are amazing organs that affect every system of the body. Their primary functions are to filter our blood and to produce urine or liquid waste. Not only do they eliminate toxins and excess fluid, but they also finely tune our body's chemistries, such as sodium and potassium levels. The kidneys are very smart; they are constantly aware of your bodily state of hydration and produce urine in amounts to maintain a proper state of hydration around the clock.

The ureter—not to be confused with urethra—is the delicate tube that carries the urine from the kidney to the bladder. It is also the tube that, when blocked by the passage of a small kidney stone, causes enough severe back pain to bring a grown man to tears.

Adrenal Gland

The adrenal gland is a small gland located on the top of each kidney. Other than being its neighbor, this gland has no direct relationship to the kidney. In addition to producing adrenaline, it releases hormones that control steroid levels, salt levels, and other bodily functions in conjunction with other hormone-producing organs. The gland also produces a small amount of sexual hormones. Interestingly, cholesterol is the major component used to manufacture adrenal hormones. (This doesn't mean you can skip your cholesterol medication when your levels are high!)

Nerves

The structures down there have a variety of nerves to control their functions. These nerves separately control erections, sexual climax, skin sensation, urinary control (continence), and bladder emptying. Damage to these nerves can cause loss of any of the above functions. Neurological conditions such as Parkinson's disease, stroke, multiple sclerosis, or trauma to the spinal cord can also hamper nerve function in serious ways.

Pituitary Gland: The Other Control Center

The brain is often referred to as the largest sex organ. Although the brain deserves most of the credit for controlling our sexual response, a small gland located just below the brain—the pituitary gland—releases hormones under the direction of the brain (via the hypothalamus). One of these hormones—luteinizing hormone (LH)—controls the production of testosterone in the testicle.

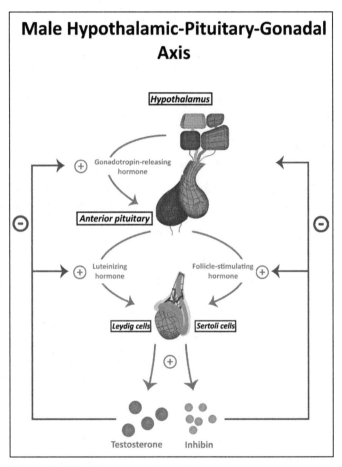

Figure 8: The Pit Bosses, the Balls

The pituitary gland produces a vast array of hormones that control a variety of bodily functions, both down there and elsewhere. Follicle stimulating hormone (FSH) stimulates the production of sperm in the testicles. Hormones from the pituitary gland also help control the function of the kidneys, adrenal glands, thyroid, and breasts.

Conclusion

Now that we know the basic makeup of everything hanging down there, let's delve into how to keep everything fine-tuned and running smoothly, as you likely maintain your automobile. That good practice for your car is even better for your body. Both are complicated machines, with many moving parts that require and deserve your care and attention.

CHAPTER 2

Benign Prostate Conditions

Peter, age fifty-six, was feeling a bit feverish over the last few hours and knew something was not right down there. He had noticed some vague urinary and groin symptoms over the previous weeks that would "come and go" (not in the desirable sense of these words). He was referred to a urologist who diagnosed him with a prostate infection.

What Is the Prostate?

Most men do not know what a prostate is—that is, not until their prostate starts causing problems. The prostate is a small gland with the sole purpose of producing a portion of the fluid expelled during ejaculation or at the time of orgasm. The remainder of this fluid—or ejaculate—is produced by the seminal vesicles (two adjacent glands) and the testicles. The prostate does not provide any other sexual function. However, this small gland is surrounded by some very important and somewhat delicate structures that are responsible for normal urinary and sexual function.

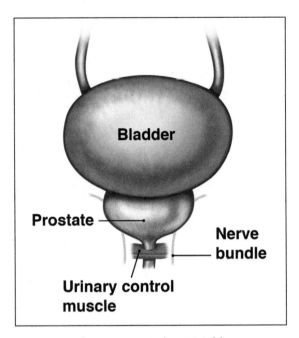

Figure 1: The Prostate and its Neighbors

The prostate resides below the bladder and just in front of the rectum. The bladder empties its contents of urine through a tube (the urethra) which passes through the center of the prostate. As a result, any problem with the prostate can cause a problem with urinary flow. Although the prostate can block urinary flow, it is the urinary sphincter muscle just below the prostate that prevents urinary leakage. In addition, the nerves responsible for erectile dysfunction travel along the undersurface of the prostate, adjacent to the rectum. Interestingly, the nerves responsible for orgasm and ejaculation are located deeper in the pelvis and are usually out of harm's way from prostate problems (and the treatments for these problems).

Prostate Enlargement—When Size Matters

The prostate increases in size with time, peaking in the late fifties. Benign prostatic hyperplasia (BPH) and bladder outlet obstruction (BOO) are other common terms for this common condition that occurs in middle-age and older men. As the prostate grows, it compresses or squeezes the urethra and impedes urinary flow. So—does size matter? Yes and no! The consistency or firmness of the prostate can also affect how well the prostate "opens up" to accommodate the flow. In Peter's case, he probably had some degree of prostate enlargement that was exacerbated by swelling from the prostate infection.

When the prostate first begins to enlarge, it usually causes no urinary symptoms. At first, this enlargement does not cause any significant obstruction or blockage of the urine flow. The bladder is strong enough to overcome mild obstruction without causing noticeable symptoms. Once the obstruction becomes more significant, the first noticeable symptom is usually frequent urination rather than a reduced flow. The explanation for this urinary frequency is that the bladder becomes less efficient at emptying, and it contracts more frequently.

Consider this analogy. If you had a pile of bricks on one side of your house and wanted to move them to the other side, it would take quite a few trips if you were to carry them by hand one at a time. If you had a wheelbarrow, it would be much more efficient to place a number of bricks in the wheelbarrow; you would make fewer trips moving the bricks. The same applies to your bladder that is trying to move a large quantity of urine; instead of a single contraction to empty the bladder, the bladder makes multiple contractions that empty small quantities of urine with each contraction.

Treatment Options

The treatment approach to prostate enlargement is initially based on how bothersome these symptoms become. In addition to increasing urinary frequency both during the day and at night, prostate enlargement can cause restricted flow, straining to urinate, difficulty starting a stream, an increased urge to urinate, and dribbling of urine after urination. These symptoms can be quantitated with the American Urological Association Symptom Score questionnaire (AUASS, see Figure 2).[1]

(Circle One Number on Each Line)	Not at All	Less Than 1 Time in 5	LessThan Half the Time	About Half the Time	More Than Half the Time	Almost Always
Over the past month or so, how often have you had a sensation of not emptying your bladder completely after you finished urinating?	0	1	2	3	4	5
During the past month or so, how often have you had to urinate again less than two hours after you finished urinating?	0	1	2	3	4	5
During the past month or so, how often have you found you stopped and started again several times when you urinated?	0	1	2	3	4	5
During the past month or so, how often have you found it difficult to postpone urination?	0	1	2	3	4	5
During the past month or so, how often have you had a weak urinary stream?	0	1	2	3	4	5
During the past month or so, how often have you had to push or strain to begin urination?	0	1	2	3	4	5
	None	1 Time	2 Times	3 Times	4 Times	5 or More Times
Over the past month, how many times per night did you most typically get up to urinate from the time you went to bed at night until the time you got up in the morning?	0	1	2	3	4	5
Add the score for each number above and write the total in the space to the right.					TOTAL:	_____
SYMPTOM SCORE: 1-7 (Mild) 8-19 (Moderate) 20-35 (Severe)						

Figure 2: AUA Prostate Symptom Score

Note that an additional question assesses level of patient "bother" (Figure 3.) This becomes important when speaking to your doctor as he or she will want to know how much discomfort or bother the symptoms are and how the symptoms are impacting your quality of life.

QUALITY OF LIFE (QOL)							
	Delighted	Pleased	Mostly Satisfied	Mixed	Mostly Dissatisfied	Unhappy	Terrible
How would you feel if you had to live with your urinary condition the way it is now, no better, no worse, for the rest of your life?	0	1	2	3	4	5	6

Figure 3: Urinary Bother Score

Lifestyle Changes

At low levels of symptoms, observation and lifestyle changes are most appropriate. For instance, reducing the amount of fluids consumed in the evening may prevent awakening so often at night to urinate. Diuretics and other medications that increase the volume of urine excreted by the kidneys, and thus delivered to the bladder, can be taken earlier in the day to reduce nighttime urination. In other cases, dietary modification such as decreasing caffeinated and alcoholic beverages—both of which can act like diuretic medications—will reduce the symptoms of prostate enlargement.

A Simple Pill May Do

When symptoms become more bothersome, a variety of medications are available to relax or shrink the prostate. Alpha-blockers are the most common type of medication to treat prostate enlargement. Examples include Flomax® (tamsulosin), Uroxatral® (alfuzosin), and Rapaflo® (silodosin). These alpha-blocker medications work by relaxing the muscles in the prostate gland, increasing the flow of urine from the bladder through the urethra to the outside of the body. Both the prostate and the bladder that connects to the urethra contain an abundance of alpha nerve receptors that, when blocked by these medications, cause the bladder opening to relax and open.

Another class of medications, 5-alpha reductase inhibitors, actually cause the prostate to shrink in size. Examples include Proscar®(finasteride) and Avodart®(dutasteride). These inhibitors tend to be more effective when the prostate is significantly enlarged, perhaps as large as a small peach or plum. Both Proscar® and Avodart® may require four to six months before the prostate decreases in size. It is not uncommon for your doctor to prescribe both alpha-blockers combined with either Proscar® or Avodart®. In fact, one drug, Jalyn®, combines both of these medications into one pill.

An abundance of prostate supplements exists. The most common supplement is saw palmetto. However, none of these supplements have been scientifically shown to be effective in comparison to a placebo (see Chapter 14.)

When Pills Just Won't Do the Job

If the medications become ineffective or poorly tolerated, a surgical option to treat the prostate enlargement can be the best course of action. A surgical approach is also the best option when the prostate enlargement causes other problems: recurrent infections, blood in the urine, bladder stones, bladder damage, kidney damage, or significant urinary retention. The traditional surgery for relieving prostate obstruction is a transurethral resection of the prostate, also referred to as a TURP.

With the patient under anesthesia, the urologist places a telescope through the urethra or the tube in the penis and into the opening of the prostate. Using a specialized electric knife, the urologist removes the inner portion of the prostate in small pieces, leaving the outer portion of the prostate intact. This would be similar to removing the fruit in the orange and leaving the peel intact. This creates a large channel through which urine can easily pass. Although this is a well-tolerated procedure, it is occasionally associated with side effects. A large percentage of patients experience retrograde ejaculation in which the ejaculate goes into the bladder rather than out of the penis during orgasm (see Chapter 5.)[2] Also, a few patients might experience bleeding, urinary incontinence, or erectile dysfunction.

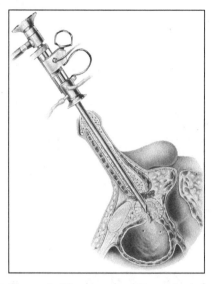

Figure 4: Transurethral Resection of the Prostate (TURP)

Several minimally invasive procedures have been developed as an alternative to a TURP in order to reduce recovery time and to decrease any possible side effects. For instance, several generations of lasers have been used to replace the electric knife used during a TURP. The most common "laser-TURP" used today is the GreenLight™ procedure.

Other types of heat can be used to destroy, shrink, or soften portions of the prostate—often with just local anesthesia and in some instances accomplished in the doctor's office. Examples include the transurethral radiofrequency ablation (TUNA®), transurethral microwave therapy (TUMT, Prolieve®), and water vapor (steam) therapy (Rezūm®). A more recent novel technique—the Uro Lift® procedure—uses small drawstring-like implants that your urologist places telescopically inside the prostate to pull open the tissue.

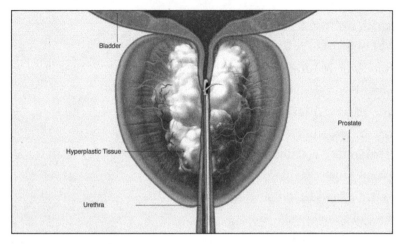

Bladder

Prostate

Hyperplastic Tissue

Urethra

Figure 5: Rezūm® Treatment

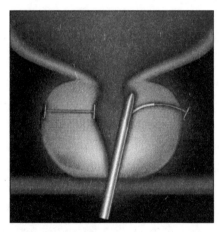

Figure 6: UroLift®

On some occasions, the prostate is too large to treat with a TURP or other alternative surgeries mentioned above. In these cases, the inner portion of the prostate or the part of the prostate that is obstructing the flow of urine is removed through an abdominal incision. This procedure—referred to as "simple prostatectomy"—is similar to prostate removal for cancer; however, for benign conditions only the inner portion is removed, thereby lessening the risk for urinary or sexual side effects.

A robotic-assisted technique is usually used for this procedure. Rather than using a large incision, the surgeon makes a one-inch incision for the insertion of a telescope and several other interchangeable instruments. The surgeon then sits at the control console a few feet from the patient, leaving the surgical assistant and scrub nurse at your side. All movements of the camera and robotic instruments are precisely performed in real time by the surgeon using ergonomic finger controls.

Aquablation® is an alternative procedure to robotic simple prostatectomy for prostates that are too large for a TURP. This technique uses a specialized telescope through the urethra that delivers a water jet to destroy the obstructing portion of the prostate. The jet is controlled by programmed robotic guidance while the urologist oversees the progress with live ultrasound imaging.

Holmium laser enucleation of the prostate (HoLEP) is yet another way to manage these large prostates through the urethra. This is a cutting laser rather than a destructive laser as used with the GreenLight™ procedure. With HoLEP, the inner portion of the prostate is separated from the outer portion and pushed into the bladder. Another instrument, the morcellator, is then inserted to cut the tissue into small pieces that can then be removed.

HoLEP techniques have a steep learning curve, thereby limiting availability of surgeons who use this procedure. Although Aquablation® and HoLEP avoid an incision, they both have a longer healing time of the resulting prostate "cavity" compared to robotic simple prostatectomy that has a much quicker "internal" recovery using a one-inch incision. With the robotic technique, the bladder is sewn inside the cavity, thereby excluding the raw surface from contact with urine and almost eliminating bleed consequences.

How to Determine Which Treatment Is Right for You

The need for and type of treatment recommended by the urologist is usually based on symptoms, but your urologist may want to perform some other tests. A urinalysis will be performed as a screening test for other potential causes such as a urinary tract infection, kidney or bladder stones, blood in the urine, undiagnosed diabetes, or (in rare cases) bladder cancer.

Although prostate cancer can mimic the same conditions mentioned in this chapter, that is rarely the case. In fact, even for someone who has been diagnosed with prostate cancer, any urinary symptoms are more likely to be related to coexisting benign prostate enlargement. However, your urologist may obtain a PSA (prostate-specific antigen) blood test to screen for prostate cancer. A baseline PSA blood test is also helpful since levels could be falsely elevated by some of the treatments for prostate enlargement (see Chapter 3.)

Although not usually necessary with mild symptoms, your urologist may also order some diagnostic tests to assess the severity of your condition. As a general rule, patients with other conditions such as diabetes, kidney disease, or neurologic disease will require more in-depth evaluation and earlier active treatment.

Among the simpler tests is a bladder ultrasound performed shortly after you empty your bladder (post-void residual, PVR.) This easy and noninvasive test will determine how completely your bladder empties—and how much it retains. Patients are often surprised by how much urine their bladder retains in the absence of severe symptoms. Unless other conditions are suspected, a CT scan, MRI, or kidney ultrasound are seldom necessary.

If the urinary frequency or urgency is thought to be a result of a bladder condition in addition to or instead of a prostate condition, urodynamic testing may be needed (see "Overactive Bladder" later in this chapter.) Urodynamic testing involves placing a very slender tube through the urethra into the bladder. The tube has a small pressure sensor at the tip. A computer then slowly fills the bladder and plots the internal bladder pressure and bladder volumes throughout filling and emptying. The computer also plots the flow rate and measures the retained volume.

Prostate Infection—Knocking on the Back Door

Prostate infections come in many forms. Since the infection usually causes the prostate to swell, symptoms can include all of those mentioned for prostate enlargement (urinary frequency both during the day and at night, restricted flow, urinary straining, difficulty starting a stream, an increased urge to urinate, and urinary dribbling). However, with infection, the symptoms usually develop over days to weeks rather than months to years as with benign enlargement. A prostate infection can also cause pain with urination (dysuria) or generalized pain in the groin, lower abdomen, testicles, lower back, and even the legs. On occasion, as in Peter's case, an infection can cause fever and chills.

Unlike benign prostate enlargement, a prostate infection always requires treatment. Evaluation is usually limited to physical examination, urinalysis, and urine culture. More involved testing as that mentioned with prostate enlargement is usually reserved for cases in which other conditions are suspected or when a patient does not respond to treatment. Since the prostate can harbor bacteria without allowing them to escape into the urine, finding the offending organism can often be a challenge. Antibiotics are usually targeted towards the most likely organism to cause urinary tract infections (or in some cases, sexually transmitted infections—see Chapter 12.) Because it is difficult for antibiotics to penetrate the prostate due to its poor blood supply, it can take four or more weeks of antibiotics to eradicate the culprit. As always, finish your antibiotics as prescribed by your physician. Our take-home message is that it is imperative that you follow the doctor's instructions and continue with the medication even after your symptoms have subsided.

In order to facilitate the antibiotic therapy, your urologist may also prescribe alpha-blockers (as used with benign prostate enlargement) to relieve symptoms and reduce the retention of any infected urine. Increased fluid intake will also help "wash out" or dilute any bacteria in the urinary tract. Anti-inflammatory medication such as ibuprofen and urinary analgesics such as pyridium can help relieve symptoms of pain and burning associated with a prostate infection.

Once the symptoms have resolved, your urologist may evaluate you for any risk factors that would lead to a prostate infection. The most common risk factor is actually benign prostate enlargement, as is probably the case for our dear Peter. When some prostate enlargement symptoms remain after completely treating the infection, your urologist may recommend continuing alpha-blockers in order to prevent recurrence of the infection.

On occasion, the infection can be difficult to eliminate fully. Sometimes the solution can be as simple as an extended course of the same or different antibiotic. When standard cultures do not reveal the offending organism, prostate secretions can be obtained using a prostate massage in order to send for culture. In the case of sexually transmitted infections, the partner must also be treated—even if your partner has no symptoms. When these measures are unsuccessful, ultrasound, CT scan, urodynamic testing, and/or cystoscopy may be needed to look for hidden causes.

When patients experience repeated and/or prolonged prostate infections, the condition can become more chronic in nature. "Chronic prostatitis" is usually caused by inflammation rather than infection—although a bacterial infection was the inciting event. As a result, seldom does chronic prostatitis respond to antibiotics. Anti-inflammatory medication (ibuprofen) and alpha-blockers are the mainstay therapy. In refractory cases, minimally invasive therapies for prostate enlargement such as laser vaporization, microwave, or UroLift have been tried. Presently, we have no scientific evidence that minimally invasive treatments are effective in treating chronic prostatitis.

Pain in the Prostate

Prostate pain syndromes can be the biggest challenge for a urologist to evaluate and treat, and the most aggravating experience for the man suffering from this malady. Many varieties of pain exist, including prostadynia (isolated prostate pain), pelvic pain syndrome, and interstitial cystitis. Evaluation begins with eliminating the conditions already discussed in this chapter. An accurate diagnosis is important since this condition seldom responds to surgical management.

Treatments for prostate pain syndromes are directed by the symptoms themselves. As mentioned earlier, alpha-blockers relax the prostate. However, in some cases, it is the muscles around the prostate that experience "spasms." For these patients, anti-spasmodics such as valium may be helpful. In other cases, a nerve stimulator (Interstim®) can be implanted in the tailbone to "distract" the nerves leading to the bladder, prostate, and pelvis. These conditions and treatments are discussed in detail in Chapter 10.

Overactive Bladder: The Prostate's Menacing Neighbor

Although the most common cause of urinary frequency in a middle-aged male is prostate enlargement, the bladder can also be the culprit. A bladder is like a crying baby. Whether a baby has a dirty diaper, is hungry, or is just tired, the cry sounds the same. Similarly, an overactive bladder, an enlarged prostate, a bladder infection, or some combination of these conditions can all lead to frequent urination. As such, in a man with an enlarged prostate whose main symptoms are urinary frequency and urgency but whose bladder does not retain significant urine, treatment might be better directed towards "calming" the bladder with anticholinergic medications (Ditropan®, Detrol®, Toviaz®, Vesicare®) or bladder-specific beta agonist medication (Merbetriq®).

However, these "bladder-calming" medications can work too well and cause your bladder to *retain* urine (especially with coexisting prostate enlargement). Therefore, your urologist may repeat the ultrasound measurement of your bladder immediately following urination (post-void residual, PVR) once you have been on this therapy for several weeks. Other possible side effects of anticholinergic medications include dry mouth, constipation, blurred vision, and occasionally confusion (particularly in the elderly). Since Myrbetriq® works through a different mechanism, these side effects can be avoided; be aware that Myrbetriq® can cause an elevation of blood pressure in a minority of patients.

So, What Happened to Peter?

Peter was placed on a common antibiotic (ciprofloxacin), Flomax, and ibuprofen. Within one week his symptoms resolved, but he completed the antibiotics as prescribed by his doctor. He no longer needed the ibuprofen after the first week, but to this day he continues using the Flomax. His flow has never been better!

Bottom Line

Prostate ailments can present in a variety of forms. As such, your doctor will evaluate your specific condition and will guide you through the best treatment options. Early diagnosis and treatment are the key to avoiding a chronic prostate inflammatory or pain syndrome.

The prostate is man's best friend during his youth and fertility years. Upon reaching middle age, it can become a gland of pain and discomfort. However, there are effective treatments so that men don't have to suffer from conditions "down there" related to their prostate gland.

References

1. Barry, Michael J., William O. Williford, Yuchiao Chang, Madeline Machi, Karen M. Jones, Elizabeth Walker-Corkery, and Herbert Lepor. "Benign Prostatic Hyperplasia Specific Health Status Measures in Clinical Research: How Much Change in the American Urological Association Symptom Index and the Benign Prostatic Hyperplasia Impact Index Is Perceptible to Patients?" *Journal of Urology* 154, no. 5 (November 1995): 1770–74. https://doi.org/10.1016/S0022-5347(01)66780-6.
2. Campbell, Meredith Fairfax, Patrick C. Walsh, Alan J. Wein, and Louis R. Kavoussi. *Campbell-Walsh Urology.* Philadelphia: Elsevier, 2016.

CHAPTER 3

Prostate Cancer

Jerry just started his new job at a large investment banking firm. At age forty-nine he was the picture of health, but he still decided to take advantage of the executive physical offered as benefit by his company. He passed with flying colors except for one abnormal blood test—a slightly elevated PSA (prostate-specific antigen) blood test. Suddenly, the possibility of prostate cancer was looming in front of him.

The company physician referred him to a urologist, Dr. Samples. In the weeks to follow, the urologist repeated the PSA test and ordered additional blood tests. Eventually, Jerry underwent a prostate biopsy. Unfortunately, the biopsy revealed prostate cancer. Dr. Samples sat down with Jerry and his wife to formulate the best treatment for Jerry.

Introduction

The prostate gland is one of those organs that most men give very little consideration until something goes wrong. Prostate issues can manifest in an array of symptoms, but the prostate can stealthily reveal no symptoms—or even confusing symptoms—when it misbehaves. The most common prostate problems are enlargement, infection, and cancer. In this chapter, we will focus on everything you need to know about prostate cancer. See Chapter 2 for information about other prostate concerns.

What Is the Prostate Gland?

The prostate gland is a small, walnut-sized, rubbery organ that is located just below the male bladder. It resides within an inch of the skin located between the scrotum and anus (the medical term for this external area is the perineum). Adding to its cryptic nature, it surrounds the urinary tube (urethra) as it exits

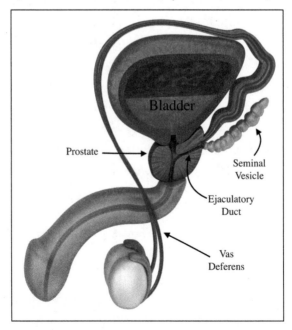

Figure 1: Ejaculatory Duct

the bladder. Since you urinate through a passageway in your prostate, any problems with this gland can significantly affect your urine flow and produce bothersome urinary symptoms.[1]

The primary function of the prostate gland is to produce some of the fluid for ejaculation that is mixed with fluid from the neighboring seminal vesicles and sperm produced in the testicles. All three of these structures join together as one common tube, the ejaculatory duct, just prior to entering the urethra.

The prostate has no other urinary or sexual function. However, it is intimately surrounded by delicate nerves and organs involved in urinary, sexual, and bowel function. Also, the prostate gland does not produce any male hormones such as testosterone. These are primarily produced by the testicles.

So, What Is Prostate Cancer?

A cancer forms when the genetic material, or DNA, inside a cell is altered in such a way that the cell can no longer follow the organizational rules of its neighboring cells. This rogue cell therefore has the capacity to invade nearby tissues, as well as faraway organs. However, this change in DNA—also referred to as a "mutation"—must be minimal enough so as to preserve its essential functions and to avoid detection by your immune system. In fact, most mutations result in the death of a cell.

Prostate cancer is the second most common cancer in men (behind skin cancer), perhaps as a result of the proportionately low blood supply in the prostate compared to other organs. This characteristic may protect it from exposure to your immune system (the "cancer police" who are responsible for removing any abnormal cancer cells before they cause trouble). Think of the prostate gland as a very crowded city packed with citizens—or in this case, prostate cells. Like any overpopulated area, the "cancer police" (or your immune system) struggles to enforce the rules. On the other hand, once a cell escapes, it must contend with your immune system. Perhaps this formidable immunologic ability helps explain why most men with prostate cancer survive the disease. Nevertheless, since prostate cancer is so common, it is the second leading cause of cancer death in men, with lung cancer being the most common cause of death in middle-age and older men.

Uncovering the "Terror Cell"

Most patients with prostate cancer do not have any symptoms at the time of diagnosis. Early prostate cancer typically has no symptoms. Once the cancer starts causing symptoms, it has probably already spread beyond the prostate gland. Therefore, screening is the most common method of detecting prostate cancer. Two tests are commonly used for this detection. One is a blood test called prostate-specific antigen, or PSA. The other is a digital rectal exam (DRE).

To Screen or Not to Screen

A lot of controversy surrounding prostate cancer screening exists. Recommendations from a variety of large organizations are all over the board, from no screening at all to screening all men yearly starting at age forty. Like most controversies, the answer probably resides somewhere in the middle.[2]

In a nutshell, concerns about aggressive screening include the risks of unnecessary additional testing, overtreatment of individuals who might do well with less or no treatment, and side effects from testing or treatment. On the other hand, in the absence of screening, most prostate cancers are discovered when a man develops symptoms. This delay most likely denies a man a reasonable chance for cure, since prostate cancer has usually escaped the prostate once symptoms develop.

So, what is the take-home message? Starting at age forty, a man should learn about his own particular risk of developing prostate cancer compared to the risks of screening.[3] Besides, this is a good time to start seeing a physician for routine health maintenance exams. For men with a family history of prostate cancer, such as a father, brother, or uncle who have been diagnosed with prostate cancer, or African American men, who have a slightly higher increase in prostate cancer than occurs in Caucasian males, early screening is probably beneficial. As a baseline, a single PSA blood test at an early age can also help determine future risk. In the absence of an alternative method for detecting prostate cancer at a curable stage, every man between the ages of forty and seventy-five should have an individualized screening plan. This plan can be created with a careful discussion with your doctor.[4]

Diagnosing Prostate Cancer

An abnormal PSA blood test or prostate exam does not necessarily mean that you have prostate cancer. This blood test can be elevated for other reasons, such as a prostate infection. Recent ejaculation and long-distance bike riding can also irritate the prostate enough to raise the PSA level. An individual man's level can vary based on age and prostate size. In order to make a diagnosis, a prostate biopsy is necessary. However, the following simple tests can help determine when PSA levels are elevated for reasons other than cancer, potentially allowing a man with an elevated total PSA to avoid a biopsy:

- Free PSA—Prostate-specific antigen is a protein that travels in the bloodstream in different forms, often attached to other proteins. One of these forms—"free PSA"—floats around unattached to other proteins. Higher free PSA levels are correlated with the likelihood of non-cancerous causes for the rise in total PSA levels. Accordingly, free PSA is sometimes referred to as "good PSA," similar to "good cholesterol" (HDL). The free PSA is often expressed as a ratio between free PSA and total PSA. This ratio should be greater than 25 percent which means that cancer of the prostate gland is less likely. If the ratio is less than 25 percent, then there is a stronger possibility of prostate cancer and additional testing may be indicated.
- Prostate Health Index (PHI)—This test combines total PSA, free PSA, and pro-PSA in order to predict the presence of prostate cancer.
- PCA3—This is a urine test that detects abnormal genetic material that can shed into the urine from the prostate immediately following a prostate massage by the urologist.
- PSA velocity—A rise in PSA of greater than 0.9 mg/dL in one year—even if the level is within a normal range—increases the suspicion of prostate cancer.
- 4K Score—This blood test examines four proteins (total PSA, free PSA, intact PSA, and human kallikrein-related peptidase 2) to differentiate men

at risk for aggressive prostate cancer from those with either a low-risk prostate cancer or no prostate cancer at all. The 4K score is useful in helping men make a decision to proceed with a prostate biopsy. A low-risk score is under 7.5 percent and a high-risk score is over 20 percent. Men with a low-risk score can be advised to defer a biopsy, and they can be safely followed since they are at low risk of adverse outcomes ten to twenty years later.

There are several variations of biopsy technique. The most common method uses ultrasound to guide the biopsy needle. A one-inch diameter ultrasound probe is gently placed in the rectum in order to visualize the prostate. A local anesthetic such as lidocaine is injected into the nerves adjacent to the prostate in order to eliminate pain during the needle biopsies. Approximately twelve biopsies are performed; you will hear a loud popping noise with each biopsy. But don't worry— its bark or (the noise that occurs with the biopsy device) is worse than its bite.

More recently, some prostate biopsies are being performed with magnetic resonance imaging (MRI) guidance. Although the MRI is more time-consuming and expensive, MRI has the advantage of potentially detecting cancers that ultrasound can miss if the cancer is small and not in the path of the biopsy needle. Another advantage of MRI is that the aggressive tumors are more likely to be seen on an MRI than the slow-growing tumors that we may not want to treat. In the past, MRI was used for men who have undergone a normal biopsy but still have a high risk of undiscovered prostate cancer (for example, progressively rising PSA.) However, the standard of care is quickly moving towards performing an MRI prior to any biopsy. This approach allows proper targeting, and it also avoids missing areas of concern and needing an MRI following the biopsy for reassurance or a repeat biopsy.

Occasionally, your doctor may recommend a general anesthesia for your biopsy. For example, sometimes a larger number of biopsies are required. Also, if the biopsy is performed through the skin rather than through the rectum, using only a local anesthetic would not be very tolerable.

Before You Solve a Problem, You Must Define It

There are two terms used to characterize any type of cancer: *stage* and *grade*. Your doctor will use these characteristics when it comes to planning treatment, prescribing medications, deciding on surgical treatments, or even estimating the likely long-term outcome or course of the disease.

Staging

Staging is a way to describe the extent or severity of an individual's cancer. As the tumor develops, it can invade nearby organs and tissues, or cells can break off and enter the bloodstream or lymphatic system. The cancer then spreads (becomes "metastatic") to form new tumors in other organs.

Your doctor will determine a cancer's "clinical stage" by using a combination of physical examination, X-ray imaging such as computerized tomography scan (CT scan) or magnetic resonance imaging (MRI scans), laboratory tests, biopsies, pathology reports, and even physical symptoms a patient describes, such as bone and back pain. If the tumor is removed, microscopic examination by the pathologist often reveals a more complete "pathologic stage."

Prostate Cancer Staging[5]

T1: tumor present, but not detectable clinically or with imaging

T1a: tumor was incidentally found in less than 5 percent of prostate tissue resected for other reasons

T1b: tumor was incidentally found in greater than 5 percent of prostate tissue resected for other reasons

T1c: tumor was found in a needle biopsy performed due to an elevated PSA blood test

T2: the tumor can be felt (palpated) on examination, but has not spread outside the prostate

T2a: the tumor is in half or less than half of one of the prostate gland's two lobes

T2b: the tumor is in more than half of one lobe, but not both T2c: the tumor is in both lobes*

T3: the tumor has spread through the prostatic capsule (if it is only partway through, it is still T2)

T3a: the tumor has spread through the capsule on one or both sides

T3b: the tumor has invaded one or both seminal vesicles

T4: the tumor has invaded other nearby structures

---

* Note that the designation "T2c" implies a tumor that is palpable in both lobes of the prostate. Tumors that are found to be bilateral on biopsy only but which are not palpable bilaterally should not be staged as T2c.

Although competing staging systems still exist for some types of cancer, the universally accepted staging system is that of the "TNM classification." The "T" describes the size of the tumor and whether it has invaded nearby tissue, the "N" describes regional lymph nodes that are involved, and the "M" describes distant metastasis (spread of cancer from one body part to another). For instance, a patient with a small prostate nodule but no evidence of spread to lymph nodes or other locations would be classified as "T2aN0M0."

Prostate Cancer Grading

Cancer grading (also called tumor grading) is a system used to classify the aggressiveness of cancer cells in terms of how abnormal they look under a microscope and how quickly the tumor is likely to grow and spread. This information can be obtained from a small sample (biopsy) of the tumor. However, as in staging, the information can be more accurate by microscopic examination of the entire tumor following complete removal. Prostate cancer has a unique grading system for classifying the aggressiveness of prostate cancer tissue: the Gleason grading system.[6]

Based on how the tumor sample looks under a microscope, a pathologist determines the level of aggressiveness on a scale from 1 to 5. Since more than one level can exist in the same prostate, the two most predominant Gleason "grades" are added together to give a Gleason "score" (for example, "3+4=7"). Since Gleason grades range from 1 to 5, Gleason scores will range from 2 to 10. These scores can indicate how quickly the tumor is likely to grow and spread. When more than one Gleason score is present, the most predominant grade is written first ("4+3=7" is more aggressive than "3+4=7").

A low Gleason score means the cancer tissue is similar to normal prostate tissue, and the tumor is less likely to spread. A high Gleason score means the cancer tissue is very different from normal, and the tumor is more likely to spread.

To make things even more complicated, pathologists no longer use Gleason scores 1 or 2. Therefore, in essence, the Gleason System scores prostate cancer cells on a scale from 6 to 10. As such, the World Health Organization is replacing the Gleason System with the new Five Group Grading System.[7] This newer system is slowly being adopted by the medical community. During this transition, both systems are usually reported in the following manner:

- Grade Group One (Gleason score ≤6)
- Grade Group Two (Gleason score 3+4=7)
- Grade Group Three (Gleason score 4+3=7)
- Grade Group Four (Gleason score 8)
- Grade Group Five (Gleason scores 9–10)

Prostate Cancer Risk Groups

Using stage, grade, and PSA levels, various methods have been proposed to categorize overall level of risk. The most widely used system was developed by the National Comprehensive Cancer Network (NCCN)[8] as follows:

- Very low risk—The tumor cannot be felt during a DRE and is not seen during imaging tests but was found during a needle biopsy (stage T1c). PSA is less than 10 ng/mL. The Gleason score is 6 or less. Cancer was found in fewer than three samples taken during a core biopsy. The cancer was found in half or less of any core.

- Low risk—The tumor has all the following characteristics:
  - Classified as stage T1a, T1b, T1c, or T2a
  - PSA less than 10 ng/mL
  - The Gleason score is 6.
- Intermediate risk—The tumor has two or more of these characteristics:
  - Classified as stage T2b or T2c
  - PSA between 10 and 20 ng/mL
  - The Gleason score is 7.
- High risk—The tumor has two or more of these characteristics:
  - Classified as stage T3a
  - PSA level higher than 20 ng/mL
  - The Gleason score is between 8 and 10.
- Very high risk—The tumor has both of the following characteristics:
  - Classified as stage T3b or T4
  - The Gleason score is 5 for the main pattern of cell growth, or more than four biopsy cores have Gleason scores between 8 and 10.

The Role of Testosterone

A lot of controversy exists as to whether testosterone causes prostate cancer. Testosterone replacement can make prostate cancer grow faster and can elevate PSA levels and prostate size in those without prostate cancer. However, there is no evidence to support that testosterone actually causes prostate cancer.

Most of these concerns stem from the fact that depletion of testosterone can help treat prostate cancer (discussed later in this chapter). However, this relationship does not necessarily mean bringing a man's testosterone from a deficient level to a normal level will cause a man to develop prostate cancer. After a successful treatment for prostate cancer, it is reasonable to replace testosterone with careful monitoring. After all, we do not intentionally deplete the testosterone in men who we think have been cured by their treatment. Therefore, why should we deny testosterone replacement in men suffering from the debilitating effects of a low testosterone level? Of course, this decision is best made through discussions with your physician.

The Role of Genomic Testing

Since the formation of cancer cells all starts with changes in a cell's genetic material—also known as DNA—we can perform specialized testing that can help further define the behavior of an individual's prostate cancer. So, what is the difference between "genomic" testing and "genetic" testing? Genetic testing looks at your genes, whereas genomic testing looks at the genes of the cancer cells. The following tests are currently the most commonly used biomarker tests that search for specific genetic mutations that can help predict a patient's prognosis:

- Prolaris®—For initial biopsy specimens, this test can help predict the chance of dying from prostate cancer within ten years. When the prostate is removed, this test can help predict the likelihood of the PSA levels rising within ten years ("biochemical recurrence" or "BCR").

- OncotypeDX®—This is a pre-operative test of the biopsy that predicts the likelihood of higher grade or escape from the prostate upon microscopic examination of the prostate following surgical removal.
- Decipher®—This test is performed on the surgically removed prostate to predict the likelihood of metastatic spread of the disease within ten years following surgery.
- ConfirmMDx®—When a biopsy does not reveal cancer but a high level of suspicion still exists, the normal biopsy material can be tested to predict the likelihood of undiagnosed cancer elsewhere in the prostate gland.

In addition to predicting who would benefit most from aggressive therapy, biomarkers can help patients' expectations and help predict outcomes of treatment. Genomic testing can even help determine the necessity of treatment in low-grade prostate cancers (especially in older men or men with other medical conditions such as heart disease, diabetes, and hypertension.)

Further testing

The PSA blood test, prostate exam, and biopsy results are the primary determinants of the extent, or "stage," of the disease. In higher risk cases, some additional testing may be recommended.

Bone scan

When prostate cancer does escape the prostate, it can travel to bone. Prior to this test, a slightly radioactive substance (radionuclide) is injected into the veins. This radionuclide then binds to prostate cancer that has migrated to the bones and is measured by a specialized camera. The radionuclide will also bind with other abnormalities of the bone such as arthritis or prior trauma. An experienced radiologist can usually differentiate these conditions from cancer.

CT Scan

Computer tomography (CT scan) is sometimes used to look for spread of prostate cancer to lymph nodes or other organs. Typically, this test is recommended for patients with a PSA level greater than 25 or high-grade disease.

MRI Scan

MRI can also be useful for treatment planning. For instance, knowing where the cancer is located relative to the nerves responsible for erectile function can help your physician preserve this ability. An MRI scan of the prostate can reveal a lot more than just its anatomy. Specialized software can distinguish the function of cancer tissue from neighboring normal tissue by measuring the rate at which fluid passes through the cells. We always want another thing to hang our hat on!

PET Scan

When searching for prostate cancer that has spread to other parts of the body, a new type of PET scan (PSMA-PET) uses a two-step process. First, a radiotracer (slightly radioactive substance) is injected into a vein; it binds to prostate cancer cells, wherever they may be hiding. Next, the resulting X-ray images are superimposed onto a CT scan to highlight any areas of suspicion. Even more mind-boggling, someday soon medications will be attached to this radiotracer to attack and kill remote prostate cancer cells like a guided missile.

What's the Answer for the Cancer?

Once the prostate cancer is detected, there may be several effective treatment options. Every person's situation is unique. Factors that influence a treatment decision include age, overall health, extent (stage) of the disease, aggressiveness (grade) of the disease, and personal preference of the patient. Nine out of ten prostate cancer cases are caught early. Most of these cases are best treated with either removal or destruction of the prostate gland. However, in some men with very early and indolent disease, careful observation ("active surveillance" or "watchful waiting") is the most appropriate choice. Patients who are diagnosed late—once the disease has spread to other locations—may benefit most from hormonal or chemotherapeutic medications.

Active Surveillance

Since prostate cancer is a slow-growing disease, patients with a small amount of low-grade disease can often be safely observed. This option is aptly named "active surveillance" since it requires regular PSA blood testing and ongoing repeat prostate biopsies. PSA alone is not reliable for determining progression of the cancer. Genomic testing and MRI scan can also enhance the ability to assess who are the best candidates for active surveillance.

You may ask, "Why would I consider leaving cancer untreated in my body?" In reality, active surveillance should actually be considered a treatment rather than "doing nothing." Just like the other treatment options, you should weigh all of the advantages and disadvantages. A proper candidate would have a low volume of cancer (three or less cancerous biopsy cores, each with less than 50 percent involvement) and a low Gleason score. Since prostate cancer is usually slow-growing, such a patient would have a wide window between the time of initial diagnosis and the time at which the likelihood of cure would decrease. In some men, this time span is indefinite. In others, treatment is merely delayed a few years, thereby avoiding exposure to the risks of treatment for a period of time.

If your urologist feels that you are a potential candidate for active surveillance, he may want to do some further testing. For instance, genomic testing can help determine how aggressive your cancer may be. In some patients, particularly those with an enlarged prostate, an MRI scan can search for an abnormality that may have slipped between the biopsy needles.

The consideration of active surveillance is very similar to a job search. Your biopsy and PSA results are your résumé. If your urologist then considers you to be a potential candidate, additional testing (genomic, MRI) serves as the interview. Even if you are offered the "job" of active surveillance, you do not need to accept. In fact, at any time you can move onto other treatment prospects.

Obviously, the advantage of active surveillance is avoidance of treatment side effects. The disadvantages of this approach are the emotional toll and the remote possibility that the cancer can escape the window of opportunity for cure. In addition, those on active surveillance must commit to frequent PSA testing and repeat prostate biopsies every one to two years.

Watchful Waiting

At the risk of being caught up in semantics, watchful waiting is very similar to active surveillance, but with a much lower intention of shifting to one of the active treatments listed below. As such, repeat biopsies are seldom performed and PSA testing is less frequent. The best candidates are men with a life expectancy of less than ten years due to other health issues and/or age. Watchful waiting focuses mainly on prevention and early detection of the complications of prostate cancer.

Surgical Removal

The most common method of prostate removal is robotic prostatectomy. Robotic prostatectomy is a type of laparoscopic surgery—with an added layer of technology. As with other laparoscopic procedures, rather than using a large incision, the surgeon makes a button-size incision in the abdominal cavity for the insertion of a telescope. After expanding the abdominal cavity with carbon dioxide gas, several additional small incisions are made to place narrow tubes used for interchangeable instruments. Instead of the surgeon's hands directly moving the instruments, the robotic device is wheeled up to the patient, and the robotic arms are attached to the telescope and the instruments.

The surgeon then sits at the control console a few feet from the patient, leaving the surgical assistant and scrub nurse at your side. One or two additional small tubes are often placed for the surgical assistant to use. The surgeon then views a highly magnified, three-dimensional image of the patient's interior structures. All movements of the camera and robotic instruments are precisely performed in real time by the surgeon using ergonomic finger controls. The tips of these instruments can make any wrist-like turn that the surgeon so desires. The procedure is performed using instruments such as miniature tweezers and scissors the size of a fingernail (although these scissors appear to be the size of hedge clippers to the surgeon observing them on the video monitor.)

Possible side effects of surgical removal include incontinence and erectile dysfunction. Your urologist can guide you on your likelihood of experiencing either of these side effects.

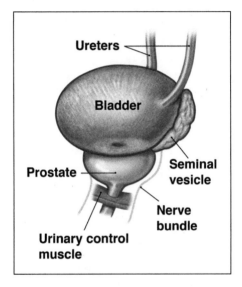

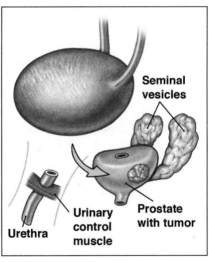

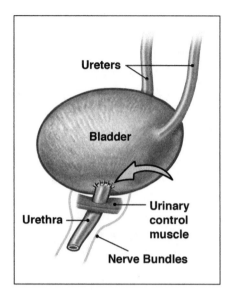

Figure 2: Prostate Removal

Radiation Therapy

There are many ways to deliver radiation energy to the prostate. The two most common ways are implantation of radioactive seeds—also called brachytherapy—and external beam radiation. With brachytherapy, the radioactive seeds are implanted using needles placed through the skin located between the scrotum and the anus (the area referred to as the perineum). A similar technique—high dose radiation, or HDR—involves placing high energy needles temporarily into the prostate and removing them prior to your leaving the medical facility. Usually, HDR is performed two separate times spaced one to two weeks apart.

Other forms of external radiation sometimes used for treating prostate cancer include proton beam therapy and stereotactic body radiation therapy (SBRT, CyberKnife®). Although these two therapies have some theoretical advantages, so far

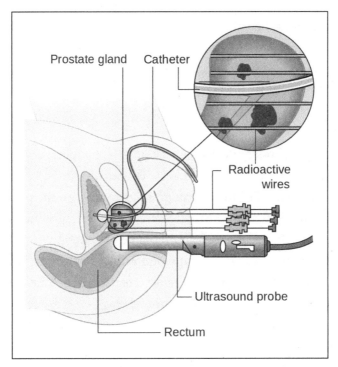

Figure 3: Prostate Seed Implantation

either has been shown to be superior to the other forms of radiation with respect to cure rates or side effects.

Possible side effects of radiation can include incontinence, radiation cystitis (painful and/or frequent urination), stricture (scar tissue formation with urinary blockage), erectile dysfunction, and radiation damage to the rectum. Years later, a slight increase in subsequent bladder, rectal, and prostate cancers may occur. Your physician can guide you on your likelihood of experiencing any of these side effects.

Freezing

Cryotherapy is a well-established technique in which the prostate is frozen to a temperature below which no cells can live.

Unfortunately, because of the high side-effect profile (especially erectile dysfunction), it is seldom used for first-line therapy. However, it is still being used for focal therapy (see below) and prostate cancer recurrence following radiation therapy.

Heating

Neither can prostate cells survive at high temperatures. High intensity focused ultrasound (HIFU) delivers directional and precise ultrasound waves through the rectum into the prostate, while visualizing the systematic destruction in real time. HIFU can potentially have a more predictable destruction of prostate tissue compared to cryotherapy and radiation therapy, thereby minimizing the possibility of side effects. This technology is seeing an increase in use for focal therapy.

Focal therapy

In a select group of men who have cancer in one relatively small area of the prostate, focal therapy can be used to destroy that one area, thereby directly minimizing the chance of urinary or sexual side effects.

HIFU, focal laser ablation (FLA), and cryotherapy have all been used in this manner. The key to success is ensuring that the cancer is isolated to the targeted portion of the prostate. An MRI of the prostate may provide this information, but another option is to perform a saturation biopsy (three to four times the number of needles compared to a standard prostate biopsy.)

Hormone Therapy

When prostate cancer escapes from the prostate to lymph nodes, bones, or other distant locations, treatment is focused on systemic therapy—medicines that treat the cancer wherever it may be lurking. Hormone therapy—more commonly referred to as androgen deprivation therapy, or ADT—has been the foundation of this treatment for many decades. The growth and survival of most prostate cancer cells are dependent on the presence of the male hormone, testosterone. In the absence of testosterone, most of the cancer cells will either die or regress. One way of achieving this goal is to remove both testicles (bilateral orchiectomy), the major source of this hormone (the adrenal glands produce one-tenth.)

More commonly, injections every one to six months are given to achieve the same effect. These medications include leuprolide (Lupron®, Eligard®), goserelin (Zoladex®), triptorelin (Trelstar®), and degarelix (Firmagon®.) Although these drugs are widely effective at controlling metastatic prostate cancer, their effect is usually temporary. The duration of success is dependent mostly on the extent and aggressiveness (Gleason score, rate of PSA rise) at the time of initiating this therapy. When these drugs fail, drugs such as abiraterone (Zytiga®) and enzalutamide (Xtandi®) can further block the stimulatory effect of male hormones on prostate growth. For more aggressive or advanced prostate cancer requiring hormone therapy, these blockers are sometimes started at the same time as the initial ADT.

Side effects of ADT include hot flashes, decreased sexual desire, erectile dysfunction, fatigue, depression, breast tenderness/growth, and osteoporosis. Some studies have shown that ADT can raise the risk of a cardiovascular event, but this risk is small—especially when compared to the risk of cancer progression. ADT is sometimes used in patients undergoing radiation therapy for cancer isolated to the prostate region, either as an initial therapy following diagnosis or following surgery. The theory behind this approach is that the medication will make the prostate cancer cells more susceptible to the radiation.

Hormone Resistance

When ADT is no longer effective at controlling the growth of prostate cancer, the condition is referred to as castrate-resistant prostate cancer, or CRPC. These unfortunate patients have a rising PSA and/or growth of tumors on X-ray, despite maintaining a low testosterone level.

Immunotherapy

The only commercially available immunotherapy for prostate cancer is Provenge® (sipuleucel-T). This therapy is also called a therapeutic vaccine. Like a preventative vaccination, it will cause your immune system to fight the disease—in this case, the prostate cancer.

The treatment involves donating blood that is then filtered by a special lab to isolate specialized cancer-fighting white blood cells. These cells are stimulated with proprietary proteins and then reinfused into your bloodstream. This process is repeated every two weeks for a total of three treatments. Interestingly, although this treatment has been shown to prolong life, the PSA values often do not decrease.

Chemotherapy

Chemotherapy is usually used in CRPC patients who either have symptoms from their cancer or when all other treatments have failed. The two most common chemotherapy drugs are docetaxel (Taxotere®) and cabazitaxel (Jevtana®). Possible side effects include hair loss, mouth sores, nausea, vomiting, diarrhea, and fatigue.

Targeted Therapy

PARP inhibitors are a type of drug that can kill resistant cancer cells in men who have a specific genetic mutation. These men lack an enzyme that repairs DNA in prostate cancer cells that are targeted by the PARP inhibitors. Examples of these genetic mutations include the BRCA gene. In fact, men with this defect are more likely to have prostate cancer and for that cancer to be more aggressive. One in four men with advanced prostate cancer have this gene. The BRCA gene also increases the risk of both prostate cancer and breast cancer in family members.

Bone Health in the Patient with Advanced Prostate Cancer

One of the major concerns with androgen deprivation therapy is loss of bone density leading to osteoporosis (or its lesser counterpart, osteopenia). In older men who may already be experiencing some bone loss, a bone density scan (not to be confused with a bone scan mentioned earlier) should be performed to measure a baseline prior to initiating ADT.

Lifestyle management is the foundation to prevention of bone complications in prostate cancer patients receiving ADT. These men should be advised to avoid smoking and excessive alcohol consumption. They should also perform regular weight-bearing aerobic and resistance exercise to stimulate bone growth and to increase muscle strength to support their skeleton. To prevent calcium and vitamin D deficiencies, they should take daily supplements of at least 1200 mg of calcium and 800 IU of vitamin D.

All men receiving ADT should have their bone density monitored at least yearly. If their bone density decreases despite the above measures, they should follow the routine guidelines for the management of osteoporosis (See Chapter 9.) In men with castrate resistant prostate cancer and bone metastases (spread to the bone), pharmaceutical intervention is mandatory since fractures are more common in this scenario. Commonly used medications include zoledronic acid (Zometa®) and denosumab (Xgeva®, Prolia®). Your physician can determine which regimen is right for you.

In men with CRPC and symptomatic (painful) bone metastases, Xofigo (radium Ra 223 dichloride) injection is given to prevent fracture and reduce the pain. This intravenous drug binds to the affected bone and delivers radiation directly to the cancerous lesions. Chemotherapy should not be given at the same time since together they can suppress the bone marrow.

Managing the Side Effects of Treatment

The best physicians are the ones who are there for you when things are not going just right. Most men who experience incontinence following surgery or radiation will recover with time and/or pelvic floor exercises (Kegels). For the remaining minority, there are several surgical solutions, all of which are discussed in Chapter 11. Unfortunately, medications seldom work in these scenarios.

Recovery of erectile function is dependent on three things: pre-treatment function, the nature of the treatment, and post-treatment rehabilitation. The more problems a man has with his erections prior to treatment, the more likely he is to lose that function. Of course, more extensive disease requires more extensive treatment which can lead to damage of the neighboring nerves that supply erectile function. In any case, encouraging blood flow as soon and as often as possible following treatment will maximize the chances of recovery. Temporary lack of blood flow to the penis can lead to permanent scar tissue formation in the microscopic vascular spaces within the penis. Aggressive rehabilitation can help prevent this process. Erectile rehab can include any combination of the following:

- Daily dosing of medications such as sildenafil (Viagra®), tadalafil (Cialis®), and vardenafil (Levitra®, Staxyn®);
- Daily use of a vacuum erection device;
- Regular use of prostaglandin PGE penile injections (Caverject®, Edex®) or urethral suppositories (MUSE®).

No better example of "use it or lose it" exists! Management of erectile dysfunction following prostate cancer treatment is discussed in detail in Chapter 4.

The side effects unique to radiation therapy can usually be managed with oral medications for urinary symptoms and topical creams for rectal symptoms.

Choosing a Treatment

Many factors go into deciding the best course of action for treating prostate cancer. These include preexisting symptoms, extent/volume of disease, Gleason score, PSA level, age, and emotional factors. Every treatment choice carries some risk, even if performed with minimal or no incisions. Whatever decision you choose, extensive experience of the treating physician is essential in order to maximize a good outcome. Of course, there is no substitute for a direct conversation with your doctor. In the end, comfort with your decision is as important as the decision itself. Usually, the first step in determining your best treatment option is considering whether you are a candidate for active surveillance. For those who conclude that they are best suited for active treatment, most will narrow the choices down to surgical removal or radiation. Choosing between these two treatments can be daunting.

The biggest advantage to radiation therapy is that it is easier to undergo than surgery. Even with robotic technology, its small incisions are not as small as the openings made by a couple of dozen needles inserted into the skin behind the scrotum when radiation seeds are implanted. General recovery from robotic prostatectomy is usually two to three weeks, whereas recovery from radioactive seed implantation is one to two days. Many patients who choose the radiation route will also undergo daily external beam treatments for four to seven weeks.

The biggest advantage to surgical removal is the information learned that is not available through other treatment methods. Once the prostate is removed, it can be fully analyzed to determine the extent, location, and grade of the disease within the prostate and seminal vesicles (and lymph nodes if necessary.)

More importantly, the ability to monitor a patient for possible recurrence is dramatically enhanced. When the prostate is removed, the PSA blood test should become undetectable (<0.1) within six weeks if all the cancer cells have been successfully eliminated. Following radiation, the PSA in some cases may never become undetectable since the prostate is still present in the body. Furthermore, even with a fully successful cure, the

PSA can dramatically fluctuate for the first several years following the radiation treatment. Although most properly selected patients can be cured with either surgery or radiation therapy, detecting the minority of treatment failures in a timely fashion is critical to selecting the appropriate back-up option. With PSA as the only guide, it can be difficult to distinguish between the expected PSA changes—or "bounce"—following radiation from surviving prostate cancer cells lurking in the shadows.

The most significant disadvantage to removal of the prostate is a higher possibility of long-lasting bladder control problems as compared to radiation therapy. The most bothersome possible disadvantage to radiation therapy involves difficulty emptying the bladder. In severe cases, patients can experience debilitating frequent urination (including multiple disruptions of sleep), decreased urinary flow, and pain with urination. Furthermore, in those patients who already experience frequent urination (day or night) or decreased urinary flow, the likelihood of long-lasting symptoms following radiation therapy increases dramatically. In fact, patients with these preexisting symptoms would most likely notice an improvement after prostate removal.

So, which of these two treatments is better at curing prostate cancer? To determine this, we would need to conduct a very large study of patients (many thousands) willing to have their treatment chosen by the flip of a coin. Since that is not possible, several theories have been proposed. With radiation, since the areas around the prostate also receive a dose, cancer cells that have just penetrated through the capsule can potentially be eliminated. However, this theory is difficult to prove since we do not know which patients have this scenario (no examination of the prostate by the pathologist). However, radiation can be given following surgery for those patients proven to have disease just outside of the prostate, thereby avoiding radiation to the surrounding areas in most patients.

Sometimes age can be a consideration in the treatment decision. By no means is the life of a forty-year-old more valuable than a seventy-year-old. However, since prostate cancer usually grows slowly, treatment failures can often be managed in older patients. In the younger patient, surgical removal provides an improved knowledge of disease status with potentially better and more timely backup options. Radiation can easily be given after surgery, but rarely is surgical removal possible following radiation. In addition, younger patients are much more likely to avoid some of the side effects associated with prostate removal. Since some of the risks of radiation can be delayed by years, these issues can be more significant for younger patients. For instance, the risk of secondary cancers of the rectum, bladder, or prostate (a new prostate cancer) can increase many years after the radiation treatment. Of course, any of the potential side effects to the bladder or rectum from radiation therapy are avoided with surgical removal.

Questions You Might Want to Ask Your Doctor

- What is my cancer stage and grade?
- Do I need any other tests?
- Am I a candidate for active surveillance?
- What are *my* chances of cure in *your* hands?
- What are *my* chances of urinary, sexual, or other side effects in *your* hands?
- What are all my treatment options?
- What do you consider the best treatment for my cancer?

- How many procedures have you performed on patients like me?
- How often do you perform these procedures and how many have you performed in total?
- Do *you* perform the entire procedure or treatment?
- What are my options should my initial choice prove unsuccessful?
- How often will I need to have follow-up exams, blood tests, and imaging tests?
- Should I have genetic counseling or testing?

So, Whatever Happened to Jerry?

Based on Jerry's biopsy, PSA, and genomic testing, Dr. Samples determined that his prostate cancer was intermediate risk. Therefore, he was not a good candidate for active surveillance. Since Jerry did not have any symptoms, he would not have known this life-saving information had he not underwent PSA blood testing. After many discussions with several physicians, he and his wife decided that robotic prostate removal was his best option. Several months after the surgery, his fear of side effects subsided when his performance in both the bathroom and the bedroom returned to the way it used to be.

Bottom Line

Prostate cancer is the second leading cause of cancer death in American men, behind lung cancer.[9] Starting at age forty, men should learn their risk of developing a harmful prostate cancer. It all starts with information easily obtained by asking the right questions and having meaningful discussions with his physician. Only then can a man make an informed decision of how he should be screened and/or treated for prostate cancer.

References

1. Campbell, Meredith Fairfax, Patrick C. Walsh, Alan J. Wein, and Louis R. Kavoussi. *Campbell-Walsh Urology*. Philadelphia: Elsevier, 2016.
2. U.S. Preventive Services Task Force. "Prostate Cancer: Screening." Final Recommendation Statement, May 8, 2018. https:https://www.uspreventiveservicestaskforce.org/uspstf/recommendation/prostate-cancer-screening.
3. ProstAware. "Educate." Accessed August 1, 2017. https://prostaware.org/educate/.
4. American Cancer Society. "American Cancer Society Recommendations for Prostate Cancer Early Detection." Last modified February 24, 2023. Accessed August 1, 2017. https://www.cancer.org/cancer/prostate-cancer/early-detection/acs-recommendations.html.
5. James D. Brierley, Mary K. Gospodarowicz, and Christian Wittekind, eds. *TNM Classification of Malignant Tumours*. 8th ed. Malden, MA: Wiley-Blackwell, 2017.
6. Gleason, D. F. "Classification of Prostatic Carcinoma." *Cancer Chemotherapy Reports* 50, no. 3 (March 1966): 125–28. https://pubmed.ncbi.nlm.nih.gov/5948714/.
7. Epstein, Jonathan I., Michael J. Zelefsky, Daniel D. Sjoberg, Joel B. Nelson, Lars Egevad, Cristina Magi-Galluzzi, Andrew J. Vickers, Anil V. Parwani, Victor E. Reuter, Samson W. Fine, James A. Eastham, Peter Wiklund, Misop Han, Chandana A. Reddy, Jay P. Ciezki, Tommy Nyberg, and Eric A. Klein. "A Contemporary Prostate Cancer Grading System: A Validated Alternative to the Gleason Score." *European Urology* 69, no. 3 (March 2016): 428–35. https://doi.org/10.1016/j.eururo.2015.06.046.
8. NCCN Foundation. *Prostate Cancer: Early Stage*. Plymouth Meeting, PA: National Comprehensive Cancer Network, 2017. https://www.nccn.org/patients/guidelines/content/PDF/prostate-early-patient.pdf.
9. Centers for Disease Control and Prevention. "An Update on Cancer Deaths in the United States." Accessed May 1, 2022. https://www.cdc.gov/cancer/dcpc/research/update-on-cancer-deaths/.

CHAPTER 4

Erectile Dysfunction

Mr. Johnson, age fifty-six, has been experiencing progressive difficulty obtaining and maintaining an erection adequate for sexual intimacy. He has a history of hypertension, elevated cholesterol levels, and adult-onset diabetes. He is fifty to seventy-five pounds overweight. He is experiencing mild depression about erectile dysfunction and has noted that his relationship with his wife of thirty years has deteriorated.

Of all the chapters in this book, this chapter will probably impact almost every male reader because impotence or erectile dysfunction (ED) affects most men during their lives. Just as Leo Tolstoy said in 1920, a tragedy in the bedroom is often devastating to a man and his partner. It impacts a man's self-confidence, masculinity, productivity in the workplace, and nearly every aspect of his life. This chapter explores the topic in detail, exposes many of the myths associated with ED, and discusses the treatment options available so that a man no longer needs to suffer the "tragedy of the bedroom."

You Are Not Alone

Erectile dysfunction (ED) is a significant problem. There are more than 150 million men in the US.; nearly 33 million will experience a failure in the bedroom. The Massachusetts Male Aging Study was probably the first study that brought to light how common erectile dysfunction is. This study demonstrated that 52 percent of all men between the ages of forty and seventy have some degree of erectile dysfunction.[1]

Although the disease is considered a benign disorder, it may dramatically impact the quality of life of many men and that of their sexual partners. ED often results in anxiety, depression, and lack of self-esteem and self-confidence, all of which can perpetuate the disorder.

A Brief History of Treating ED

Sex and intercourse are basic human needs, familiar to all people from the beginning of recorded history until today. Erectile dysfunction is by no means a recent development in medical history, nor is it one that's been kept quiet or away from public discussion. The failure of men to "rise to the occasion" has been a topic of interest to men *and* women since the dawn of human history.

Ancient Egyptians had a rich and varied sexual life chronicled in numerous hieroglyphs and papyri. In the times of the pharaohs, the Egyptians defined impotence and recorded methods of treating this disorder. Egyptian physicians used magical spells and incantations to improve the sexual prowess of their patients. Perhaps this was the first written evidence of treating erectile dysfunction having a psychological origin rather than a physical one.

Hippocrates, the father of modern medicine, wrote in one of his texts, *The Aphorisms*, "[W]hen two, three, or even more attempts are attended with no better success, they think they have sinned against the God and attribute thereto the cause."

Means of stimulating sexual desire in men were addressed in Ancient Greece and Rome. Pliny the Elder's *Natural History* advocates various root vegetables and plants as potent aphrodisiacs.

The ancient Chinese correctly correlated aging with ED: "At forty-eight, a male's potency is reduced or exhausted. As his genital secretions are exhausted, he can no longer hold it erect and, at this point, he cannot generate any more."

Ancient India also had solutions for ED. Potions consisting of the testicles of a goat cooked in milk were purported "to make a man as powerful as a bull." The Kama Sutra is the earliest example of a love-making manual in the history of the world.

Of course, there were advances from these ancient times to the Renaissance, where the famous drawings of the anatomical structures in the male and female pelvis were demonstrated to the medical community.[2]

One of the most significant and recent discoveries in the fight against ED was that phosphodiesterase-5 inhibitors increase blood supply to the penis. This resulted in the development of sildenafil or Viagra, which has made it possible for millions of men worldwide to solve their ED dilemma.

No book on ED would be complete without the mention of Alfred Kinsey, William Masters, and Virginia Johnson. They conducted groundbreaking research on human sexuality and clearly distinguished between psychological and physical cause of impotence.

Thousands of studies and reports on treatments for ED have been conducted and reported over the last two millennia. Most of these are based on the discoveries of innovative and creative physicians and scientists who made it possible for millions of men to be able to successfully engage in sexual intimacy.

Definition of ED

Erectile dysfunction is the persistent and consistent failure to achieve and maintain an erection adequate for the sexual needs of a man and his partner. An occasional failure to achieve an erection happens to most men during their lives. However, when it happens most of the time, it becomes a problem that affects a man's masculinity, erodes his confidence, and even impacts his performance and productivity in the workplace. The good news is that nearly every man with ED can be successfully treated for this common medical problem.

How Does an Erection Occur?

An erection is like filling a tire using an air compressor. To inflate a tire, you need to have the air compressor plugged into the wall and have an electric current flow to the air compressor. The air compressor is turned on. It takes air from the environment, and the air flows through the tube from the air compressor to the tire. The tubing must be sufficiently open and without obstruction or kinks that could impede the airflow. The tube must be firmly attached to the nipple on the tire, and the tire must be a closed system without any holes that could allow the air to escape.

The same principles of filling a tire with air hold true for a man's erection. The erection starts in the brain that sends a message via nerves in the spinal cord to open the blood vessels that allow blood to flow into the penis. The heart must be working properly to pump the blood into the blood vessels that supply the penis. When more blood flows into the penis through the arteries than leaves the penis through the veins, an erection will occur. Failure of any component from the brain to the end organ, the penis, will result in erectile dysfunction.

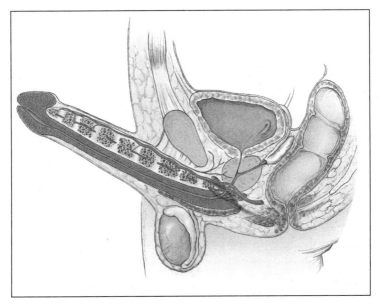

Figure 1: Erect Penis *(Coloplast)*

Causes of ED

ED can broadly be divided into two categories: psychological and physical. In the past, most ED was thought to be psychological in nature. We have identified many physical causes, such as injury to nerves and conditions affecting the blood vessels, as the most common causes of ED. For men over fifty years, physical causes are the most common. In younger men, especially under forty years, psychological causes prevail.

Physical Causes of ED

Of the physical causes, vascular disease is the most common. Arteriosclerosis, or hardening and narrowing of the blood vessels that supply the penis, causes more than 50 percent of the cases of men with ED. Increased cholesterol levels and hypertension or high blood pressure can also narrow the blood vessels and decrease the blood flow into the penis, causing ED.

Diabetes affects the nerves and the blood vessels, resulting in ED in nearly 50 percent of men with diabetes. Neurologic causes, which impair the nerves to the penis or affect the brain and spinal cord, include spinal cord injury, multiple sclerosis, Parkinson's disease, stroke, and Alzheimer's disease. Significant pelvic trauma (such as a fracture of the pelvis), pelvic surgery (such as removing the entire prostate gland for prostate cancer), and bladder or colon surgery can result in ED. Radiation therapy to the prostate gland or bladder can result in injuries to the nerves and blood supply, impacting a man's ability to have an erection. Peyronie's disease is a condition that results in the formation of a scar in the penis that can result in curvature of the penis that alters the blood supply, thereby resulting in ED.

Medications are also a culprit. Hundreds of medications can cause ED, such as blood pressure medications, heart medications, antidepressants, tranquilizers, pain medications, and sedatives, just to name a few. Table 1 lists some of the most common culprits or medications that have side effects of ED. The male hormone testosterone, produced in the testicles, is primarily responsible for libido (or sex drive), and testosterone deficiencies can affect a man's erection.

Venous leak is a condition that results in blood leaving the penis as fast as it enters and is usually associated with failure to maintain the erection once it has been achieved. It is akin to

Table 1
Drugs that cause ED

Antihypertensives (clonidine, reserpine, beta-blockers, verapamil, methodopa)

Diuretics (thiazides and spironolactone)

Cardiac/Circulatory (clofibrate, gemfibrozil, digoxin)

Tranquilizers (phenothiazines, butyrophenones)

Anticholinergics (disopyramide and anticonvulsants)
Antidepressants (tricyclic antidepressants, AAOIs, lithium, and SSRIs)

Hormones (estrogen, corticosteroids, cyproterone acetate, 5-Alpha reductase inhibitors, LHRH agonists and antagonists)

H2 antagonists (cimetidine and ranitidine)

Cytoxic agents (cyclophosphamide, methotraxate, Roferon-A)

Table 1: Medications with side effects of erectile dysfunction.

the metaphor of filling the tire air compressor while there is a hole in the tire. As fast as air enters the tire, it leaves through the hole, and the tire remains flat. This is the same problem as a venous leak in a man with ED; blood enters the penis but leaves just as quickly through the abnormal veins.

Miscellaneous causes include smoking (which affects the blood vessels), use of illicit drugs such as cocaine, and excessive alcohol consumption. Shakespeare made mention of this in *Macbeth*: "Alcohol provokes, and unprovokes; it provokes the desire, but it takes away the performance." Chronic or longstanding liver and kidney disease can impact a man's erection. Other hormonal causes include too much or too little thyroid hormone and an elevation of prolactin levels, decreasing testosterone levels. Finally, obesity is also associated with an increased incidence of ED, possibly because of the association of obesity with diabetes, hypertension, and high cholesterol levels, which are all comorbid conditions associated with ED.

Cannabinoid receptors are present in the penile tissue, so it is possible for tetrahydrocannabinol (THC), the active ingredient in marijuana, to impair penile function. In some men, this can lead to ED. It is important to mention marijuana as a potential cause of ED since it is widely available for medicinal and recreational purposes.

Marijuana can cause euphoria followed by drowsiness and slowed reaction time. Some men may have decreased desire for sex after smoking marijuana. Also, marijuana increases the levels of dopamine in the body. Dopamine is a hormone that can affect mood and sexual feelings. Once a man becomes accustomed to these high levels of dopamine, he may find that his body's natural level of dopamine may not be enough to stimulate him sexually. Marijuana can also affect the circulatory system and cause blood pressure and heart rate to increase. Both medical conditions are risks for ED.[3]

Another possible cause of ED is excessive or improper bicycle riding. When riding a bicycle, a man's weight presses on the perineum, the area between the rectum and the scrotum. The perineum is the location of the arteries and nerves that supply the penis. Since the arteries are unprotected, they're prone to damage from constant pressure from the bike seat. When a man sits on a bicycle seat, he's putting his entire body weight on the artery that supplies the penis. When the arteries are compressed, there is a decrease in the tissue's oxygen to properly function.

Prolonged or excessive bike riding can injure the blood supply to the penis, making erections difficult or impossible. You can anticipate injury if you experience numbness in the genital area after riding. The longer the numbness lasts is usually a barometer of the extent of damage you are creating for those vital blood vessels that supply the penis.

For men who are experiencing ED associated with bicycle riding, we recommend the following:

- Penile numbness and excessive genital shrinkage are warning signs that there may be too much pressure on your perineum. The nerves in the perineum are compressed, which means the artery that feeds the penis is also being compressed.

- Make the following changes in your riding style and/or your position on the bike: 1) Make sure your saddle is level, or point the nose a few degrees downward. 2) Check to see that your legs are not fully extended at the bottom of the pedal stroke. Your knees should be slightly bent to support more of your weight. 3) Stand up every ten minutes to encourage blood flow to the penis.

- Survey the multitude of anatomic racing saddles on the market, ranging from ones with a flexible nose to models with a hole in the middle. You may want to experiment with a broader, more heavily padded brand or a "double bun seat" that places the weight on the bones and off the perineum.

- Heavier riders may be more at risk of arterial compression damage because of the greater weight placed on the perineum, the area between the scrotum and the rectum. You should consider a wider saddle with extra padding if you're in this category.

- When riding a stationary bike, the tendency is to stay seated and grind against the larger gears for long periods. Get off the seat as frequently as you would on your regular bike and be sure that it's set up with the same riding position.
- Get off the seat when riding over rough or irregular terrain. Use your legs as shock absorbers.

Our take-home message on bicycle riding: Most men are not aware of the relationship between their bike and their erections. Our final advice for good health is that men shouldn't necessarily ride farther, but ride smarter.

Psychological Causes of ED

Psychological causes, which only cause ED in 10 percent of older men, include depression and recent emotional trauma, such as loss of a partner, divorce, stress, or job loss. Psychological causes can also occur with deterioration in the relationship between a man and his partner or if there are problems and distractions with other family members.

A common situation in young men is "performance anxiety." This situation occurs when a young man is anxious about his ability to perform, especially with a new partner. The ED in this scenario becomes a cyclical event as the erectile failure itself creates more anxiety with each subsequent desire to be sexually intimate.

Priapism: The Erection That Will Not Quit

We know it sounds like a dream come true when you hear about the erection that might last for four hours after taking one of the oral medications for ED. Nothing could be further from the truth! This situation, priapism, is a medical emergency that needs immediate medical attention. If you don't get treatment, it can lead to permanent ED.

Two Main Types of Priapism

Low-flow or *ischemic priapism* happens when blood gets trapped in the erection chambers. Most of the time, there's no apparent cause, but it may affect men with blood disorders such as sickle-cell disease or leukemia.

High-flow or *non-ischemic priapism* often happens after an injury to the penis or the perineum. A rupture of an artery that supplies the erection chambers or corporal bodies of the penis will result in an extended and painful erection.

The most common cause of priapism occurs following medication injections, such as trimix, prostaglandin, phentolamine, or papaverine into the penis (read about self-injection therapy, under Second-Line Treatments). Drugs that may cause priapism include medications used to treat depression and mental illnesses, such as trazodone (Desyrel) or chlorpromazine (Thorazine). Priapism can also occur following street drugs such as heroin and cocaine.

The goal of treatment for priapism is to make the erection go away to prevent permanent ED. Treatment may consist of removing the blood trapped in the penis. For low-flow priapism, your doctor can inject alpha-agonists drugs into your penis to reduce the blood flow to the penis. This may reduce the pain and swelling. Another successful treatment option is terbutaline—five milligrams, taken orally. The take-home message is that an erection lasting four hours is not a dream come true; it's a nightmare that needs immediate medical attention.

Penis Fracture

We know that most fractures involve bones in the upper and lower extremities. However, other fractures can be just as devastating, including fracturing the penis, even though there are no bones made of calcium in the penis. (The only mammal with a bone in the penis is the whale.)

Penile fractures happen only when you have an erection. When you're soft, the pressure inside your penis is low, so it's more able to bend and withstand unexpected forces. However, when the penis is erect and hits a hard structure such as the woman's pubic bone, the penis cannot bend easily, leading to large pressure increases in the penis and a blowout or fracture.

In most cases, there is a loud popping sound when the fracture occurs, and it is usually heard by both partners. This is followed by intense pain and then swelling of the penis.

A fracture of the penis is a medical emergency that requires immediate attention and surgery to repair the fracture.

Our take-home message: ED is a common problem affecting millions of American men. You are not alone. Although ED is more common in middle age and older men, it does not need to be a consequence of aging. Older men who are healthy, have an available partner, and are interested in sexual intimacy can be helped and successfully engage in intimacy.

How is ED diagnosed?

Your doctor will begin by taking a careful history. He may ask: How long have you had ED? What other diseases or conditions do you have? What medications, including over-the-counter medications or supplements, do you take? What treatments have you tried in the past? What is the impact of ED on your partner?

Next is a physical examination, which includes taking your blood pressure, examining the penis and testicles, and a rectal exam to evaluate your prostate gland. The physical examination includes evaluating the nerve supply to the lower extremities and the blood vessels in the legs and feet.

You will probably be sent for a urinalysis and a few blood tests, including testosterone level, cholesterol level, blood sugar, liver, kidney, and thyroid function tests. The urine examination is for protein and glucose (sugar), indicating kidney disease or diabetes. A prolactin level and other blood tests associated with testosterone secretion may be obtained if the testosterone level is low.

The duplex Doppler study is usually accompanied by an injection of a drug or drugs into the penis to stimulate the blood flow to the penis. Usually, these injections create an erection that lasts about twenty minutes and allows the doctor to see if the blood vessels respond appropriately to the medication. Additional tests may also include a penile color duplex ultrasound, which is used to evaluate the blood flow to the penis.

A nocturnal penile tumescence test is only occasionally ordered to differentiate psychological causes from physical causes. Most normal men have three to five involuntary nighttime erections that can be recorded with a band placed loosely around the penis and attached to a monitoring device. If a man obtains an erection, there will be an increase in the girth of the penis that is recorded on a computer monitor. The presence of normal nighttime erections indicates that the nerve and blood supply to the penis are intact and that the problem is probably related to psychological causes rather than a physical cause.

Another option for the nocturnal tumescence test can be easily accomplished by using a few US postage stamps placed around the base of the penis before going to sleep. Suppose a man has a normal nocturnal erection. In that case, there will be an increase in the girth of the penis, which will break the perforations between the stamps, indicating that the man has an adequate erection.

How Is ED Treated?

First, the good news: Nearly everyone with ED can be helped. There are many treatment options available to all men who suffer from ED. You need to find one that meets your situation and your pocketbook. The best patients are those who are aware of the options and then, with the help of their doctors, select treatments that work best for them. Most men start with the least invasive treatment options and then proceed to additional treatments if the less invasive treatments are ineffective.

Perhaps the easiest and least expensive treatment is lifestyle modification. Men who are smokers will find that there will be an improvement in their erections when they stop smoking cigarettes. The active ingredient in cigarettes, nicotine, decreases the blood flow to the penis. Prolonged smoking over many years may permanently narrow the blood supply to the penis, which will preclude the penis from becoming rigid enough for vaginal penetration.

Men who are overweight will notice an improvement in their ability to engage in sexual intimacy by losing weight and beginning a physical exercise program. For men with high cholesterol levels, dietary modification may decrease the cholesterol levels and improve the blood supply to the penis, resulting in better erections.

A question we often ask our overweight patients is the following: Mr. Smith, would you like to take a pill that improves your energy level, increases your sex drive, decreases your blood pressure, decreases your cholesterol level, decreases your risk of prostate and colon cancer, enhances your muscle mass, improves your mood, lifts your depression, and makes your penis one to one-and-a-half inches longer? Every man says, "Yes, I'd like that pill!" We respond, "Mr. Smith, it isn't a pill. It's exercise!" Yes, exercise, improvement in nutrition, and weight loss will do all the above. The loss of the abdominal protuberance will give the *appearance* that the penis is longer.

Excessive use of alcohol can impact a man's erection, and decreasing alcohol consumption will significantly improve his sexual function. Alcohol in small amounts, one to two drinks per day, may serve as a social lubricant and a sexual facilitator. Excessive alcohol may result in difficulty obtaining and holding an erection adequate for sexual intimacy. Also, excessive alcohol consumption can result in delayed orgasm or even the inability to have an orgasm, which can be a source of frustration for both the man and his partner.

The use of alcohol over many years can result in permanent liver damage. This can create hormone imbalance with an overproduction of the female hormone estrogen and a decrease in testosterone production. Both situations lead to a decrease in libido and sexual performance.

Men who use medications with the side effects of ED or decreased sex drive can speak to their doctor about the adverse side effects. The doctor may then reduce the medication dosage or change to another class of drugs that do not have the side effect of ED or a decrease in sex drive.

For example, some antidepressants have the side effect of ED. If ED occurs using an antidepressant, another medication may be tried. For instance, Wellbutrin is less likely to cause ED and can still be effective for treating depression.

If lifestyle modifications are ineffective, treatments can be organized into first-, second-, and third-line treatments.

First-Line Treatments

First-line treatments begin with oral medications. These medications are the phosphodiesterase-5 inhibitors, which consist of sildenafil (Viagra), tadalafil (Cialis), Staxyn, and vardenafil (Levitra). The FDA first approved Viagra in 1998. These drugs are indicated in men who have ED and no contraindications to their use. The main contraindication is the concurrent use of nitrate therapy for men with cardiovascular disease. This also includes men who have heart disease and carry a small container of nitrates (sublingual nitroglycerin).

Phosphodiesterase-5 inhibitors dilate the blood vessels to the penis, but they also dilate blood vessels elsewhere in the body. Nitroglycerin also dilates the blood vessels. Men who are using the ED drugs and take a nitroglycerin tablet may have excessive dilation of the blood vessels in the body. This dilation may result in a decrease in the blood supply to the heart.

Suppose a man has chest pain due to poor blood flow to the heart. The combination of medications will rob the heart of the necessary oxygen for the normal functioning of the heart. It would be like using an air compressor to simultaneously fill all four tires of a car simultaneously. It can't be done. Remember the movie *Something's Gotta Give?* Harry Sanborn, played by Jack Nicholson, takes nitroglycerin after using Viagra and is rushed to the emergency room. Using phosphodiesterase-5 inhibitors and nitroglycerin can result in a dangerous decrease in the blood supply to the heart.

We issue a precaution about using any oral medications for treating ED in conjunction with alpha-blockers such as Rapaflo, Flomax, or Uroxatral. These medications treat an enlarged prostate (see Chapter 2). We recommend men who use alpha-blockers and ED medication to *avoid taking them simultaneously.* For example, suppose a man regularly has sexual intimacy in the morning. He should use the ED drug in the morning and the alpha-blocker in the evening.

We don't prefer one first-line drug over another, as they all have a similar mechanism of action. Viagra lasts four to eight hours after ingestion. The drug should be taken on as much of an empty stomach as possible. A full stomach will delay its absorption and take longer to become effective. Cialis lasts thirty to thirty-six hours after ingestion, and Levitra lasts four to eight hours but appears to have a shorter period to start working.

All three drugs require genital stimulation for a man to achieve an erection. A man who plans to have sexual intimacy late at night may take the ED drug in the early evening before going out for dinner. He will be ready later to successfully engage in sexual intimacy. Table 2 summarizes the three most popular oral drugs used for treating ED, listing the dosage and side effects for each one.

Our experience is for one of the phosphodiesterase-five inhibitors to be effective for several months or even years and then ineffective in producing an erection adequate for sexual

| Table 2 | | |
Dosage and side effects of phosphodiesterase-5 inhibitors		
Viagra (sildenafil)	25, 50, 100mg	headaches, flushing, visual disturbances, nasal congestion, GERD (gastroesophogeal reflux)
Levitra (vardenafil)	10, 20mg	headache, flushing.
Cialis (tadalafil)	5, 10, 20mg	headache, back pain, muscle cramps

intimacy. We have found changing to another phosphodiesterase-five inhibitor may have a beneficial effect.

Second-Line Treatments

When the phosphodiesterase-5 inhibitors are no longer effective, the next treatment options consist of a vacuum device, injection therapy, or suppositories inserted into the end of the penis.

A vacuum erection device is a mechanical, nonsurgical method of filling the penis with blood, thus creating an erection that is hard enough to achieve penetration. It is based on the principle that placing the penis in a vacuum chamber or plastic cylinder can produce an erection. Air is then removed from the cylinder by either a manual or electric pump creating a vacuum or negative pressure around the penis. When the penis becomes engorged with blood, a tight elastic or rubber band is released from the plastic cylinder onto the base of the penis, thus trapping the additional blood in the penis. At this time, the man can begin to engage in sexual intimacy.

We recommend that the elastic or rubber band be left on the penis for thirty minutes or less. Most men cannot ejaculate as the semen is trapped behind the constricting band in the urethra. Although no fluid exits the penis, the constriction band will not prevent the sensation of climax. The side effects include:

- Mild pain when first used;
- Transient bruising of the skin on the penis;
- Numbness of the penis is relieved when the constricting band is removed.

Vacuum devices are often recommended for men after prostate cancer surgery, where the entire prostate gland is removed. This allows for blood to artificially be brought into the penis after surgery and is believed to reduce shortening of the penis that often occurs after surgical removal of the entire prostate gland.

Self-injection therapy is also an option when the oral medications are not effective. This is an effective method of producing an erection firm enough for penetration by injecting a small amount of medication directly into the shaft of the penis and the underlying corporal bodies (erection chambers). The medication is often trimix, a combination of alprostadil, papaverine, and phentolamine; however, an injection of only prostaglandin is available. A small amount of this medication is injected directly into the corporal bodies fifteen to twenty minutes before engaging in sexual intimacy, using a tiny needle that causes minimal discomfort. The proper trimix dosage is determined using test injections under doctor supervision. An erection will usually occur within ten minutes after the injection. The erection will be indistinguishable from a natural erection. An erection will usually last twenty to thirty minutes. Efficacy rates have been reported as high as 70 percent. It is often effective for men who do not respond to oral drug treatment.

The side effects usually consist of mild penile pain at the injection site in about 25 percent of men. This mild pain usually subsides in a day or two. Most men do not complain of the pain, and only a few men stop using these injections because of this side effect.

We advise men to limit injections to two to three times per week and never more than once every twenty-four hours. We also recommend using different injection sites on the penis. If the trimix is injected at the same place, there is a risk of scar formation and the development of Peyronie's disease.

The most significant problem with self-injections of trimix is the development of priapism. Priapism is the occurrence of a prolonged erection in the absence of sexual excitement. Priapism lasting more than four hours is a medical emergency. The man with a sustained or prolonged erection must go immediately to a doctor or the emergency room. There are medications, phenylephrine or adrenaline (epinephrine), which can be injected into the penis to reverse the priapism and allow the penis to become soft or flaccid once again. If the antidote injections for priapism are unsuccessful, an emergency surgical referral is required.

Platelet-rich plasma (PRP) is a novel treatment for ED. Platelets play a crucial role in our body's inflammatory response, tissue remodeling, and angiogenesis (formation of new blood vessels). Platelet-rich plasma is an emerging treatment for ED. This requires obtaining a blood sample from the patient. This blood sample is processed so only platelets and plasma proteins remain. These plasma proteins consist of growth factors essential for tissue restoration. The platelet-rich plasma is injected directly into the corporal bodies of the penis.

Using PRP for ED has no research to clearly show this procedure's benefits, safety, and risks. The treatment is not approved by the FDA, and at the present time it is considered an investigational treatment.

Another second-line treatment involves inserting a tiny drug-carrying pellet, called MUSE (Medicated Urethral System for Erection) into the opening of the urethra, the tube that transports urine through the penis from the bladder to the outside of the body. This opening is located on the head of the penis. This pellet, about the size of a grain of rice, is inserted into the urethra. The drug is absorbed into the surrounding tissues of the penis to produce an erection. The erection will usually occur in seven to ten minutes. The active ingredient in the pellet is alprostadil, which is also used for injection into the penis. MUSE avoids using a needle and syringe and has minimal pain or discomfort with its use. Men will usually get an erection that will last for forty-five to sixty minutes. The actual duration will vary from patient to patient. The pellets can be used up to two times a day. The drug is effective in 65 percent of the men who have used this treatment.

The most common side effect is a dull ache at the tip of the penis, which occurs in only a few of the men that use the pellet. Most men report that the discomfort lessens with repeated use of the pellet. Rarely will the erection last several hours and result in priapism which is a medical emergency. Priapism requires using other medications that will reverse the effect of the MUSE. If the man has a partner in the childbearing age group, we suggest that he use a condom to prevent the drug from being absorbed by the female partner.

MUSE comes in four strengths: 125, 250, 500, and 1,000 micrograms. The doctor selects the correct dose of the MUSE to insert that will produce the desired duration and quality of an erection and avoid any side effects or complications.

Third-Line Treatments

When first-line and second-line treatments are ineffective, surgeries become the third-line option.

A penile prosthesis is the surgical insertion of a device through a two-inch incision, allowing a man to create an erection for as long as he wants. The most popular is the inflatable penile prosthesis, developed in 1973. The inflatable prosthesis consists of two cylinders inserted into the penis, a small pump inserted into the scrotum, and a reservoir containing a harmless salt solution placed behind the abdominal muscles.

When the man desires an erection, he squeezes the pump located in the scrotum. The fluid in the reservoir is moved into

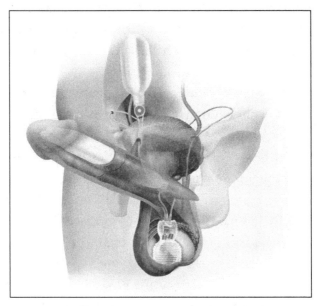

Figure 2: Inflatable Penile Prosthesis *(Coloplast)*

the cylinders in the penis. This allows the penis to increase in girth and length. This produces a natural erection.

After engaging in sexual intimacy and when he desires the penis to become soft, the man compresses a release valve located on the pump. The fluid reverses its direction and returns to the reservoir, causing the penis to become flaccid. The entire device is completely concealed and cannot be detected even under the closest scrutiny. Men with the prosthesis can comfortably shower and change clothes in the men's locker room without anyone being able to suspect that a prosthesis is in place.

The complications of the inflatable prosthesis include bleeding, infection, and failure. Rarely is bleeding or infection a problem. The device is successful in 98 percent of patients and seldom must be repaired or replaced. The patient and the partner have nearly 100 percent satisfaction with the device. The procedure is done in the hospital or a one-day surgical center. Most men are discharged within a few hours after the procedure or after they can urinate. Most men can begin using their prosthesis within four weeks after the procedure. Most insurance companies, including Medicare, pay for the surgery.

A malleable implant consists of two semi-rigid rods surgically inserted through a small incision into the penis. The man can place the penis into the position for intercourse or push it down to conceal the penis in his underwear.

In cases when vascular reconstructive surgery is performed to improve the blood supply of the penis, blocked arteries are bypassed by transferring an artery from an abdominal muscle to a penile artery. Only a small number of men would benefit from revascularization. It is not commonly performed because of the level of difficulty, high cost, and suboptimal success rate.

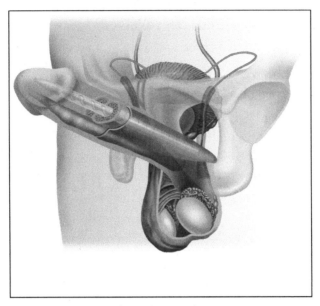

Figure 3: Malleable Implant *(Coloplast)*

It may be suggested to men who have experienced some trauma around the scrotum and anus regions.

Peyronie's Disease (PD)

PD is a health issue in which a scar or plaque forms under the skin of the penis. Most of this plaque builds up, leading to a curved erection, making intercourse difficult or painful for a man and his partner. PD is thought to impact about 6 percent of men between the ages of forty and seventy. This may be an underestimate, as many men are too embarrassed to seek help or report the problem to their physician.

Doctors do not know what causes PD or how to prevent the disease before it starts. We do know that PD runs in the family. We also know that vigorous sexual activity causes minor injuries to the penis that may initiate the process of scar formation.

The incidence is also slightly higher in men who have undergone surgery or radiation to treat prostate cancer. The disease is not related to a sexually transmitted disease or cancer of the penis.

The primary symptom of PD is curvature of the penis. A man may also feel one or more hard lumps on the penis. Men may also experience painful erections, and the man's partner may also complain of vaginal pain or discomfort. Peyronie's disease may also lead to erectile dysfunction for several reasons: blockage of blood flow by the plaque, pain from the scar tissue, or the emotional effect of the curvature.

The treatment options for PD include watchful waiting, nonsurgical treatments, and surgery. Watchful waiting is appropriate for a man with a minimal curvature that does not cause pain or impede sexual intimacy. Injection therapy includes interferon, verapamil, or collagenase drugs (Xiaflex) directly into the plaque, which may dissolve the plaque.

Surgical options include straightening surgery, referred to as penile plication. The plaque can also be removed, followed by repairing the resulting defect with a human tissue graft or synthetic tissue. Finally, the plaque can be surgically removed and a penile prosthesis inserted, thereby removing the offending bend in the penis.

Peyronie's disease is a common urologic disorder. It is a source of potential pain and discomfort. Help is available, and most men can achieve relief from the disabling curvature.

Alternative Treatment Options

Some men will opt not to use medications or consider injections or surgery to solve their ED problem. Alternative options include hypnosis, acupuncture, meditation or mindfulness, ginseng, and DHEA supplements.

Acupuncture is an ancient Chinese practice. An acupuncturist inserts six to ten tiny needles into specific sites along the body's channels or "meridians." Treatment generally occurs once a week for ten to twenty weeks.

Meditation and mindfulness are techniques of training the mind to focus on the present. Meditation reduces stress and achieves a sense of inner peace that can benefit both emotional health and overall physical condition. Types of meditation include deep breathing, visualized breathing, progressive muscle relaxation, and guided imagery. In treating ED, meditation is usually most effective when the condition is not caused by a physical issue that restricts the flow of blood to the penis but is psychological in origin. Chapter 17 discusses mindfulness and meditation and offers exercises that instill inner peace and calm.

Ginseng (or Panax ginseng) is sometimes referred to as "herbal Viagra." The root of the ginseng plant has been used in traditional Chinese medicine for thousands of years. Red ginseng contains natural antioxidants known to reduce inflammation, boost energy levels, and help the body deal with fatigue. These antioxidants may also improve blood flow through the body, including blood flow to the penis.

DHEA, or dehydroepiandrosterone, is the most common circulating steroid hormone in the human body. Men with ED often have low levels of DHEA, so, logically, taking DHEA supplements might protect against ED. Some people even believe that DHEA is a "fountain of youth" and can slow down aging.

Alternative treatments are generally safe, inexpensive, and do not worsen the ED. Remember that most alternative therapies are not covered by insurance. Most importantly, ED may indicate an underlying medical condition, so we recommend a complete history, physical examination, and a few laboratory tests.

Another alternative treatment option is low-intensity shock wave therapy. Shock waves were first used in the 1980s to treat kidney stones. This same energy source enhances the blood supply to the penis and helps restore erections.[4] Treatment consists of twelve weeks of shock wave sessions, after which a man with ED ideally doesn't need to worry about any treatment for at least two years.[5] The use of shock waves for the treatment of ED is not FDA approved. Low-intensity extracorporeal shock wave therapy should be considered investigational for men with ED.

Treatment for Psychogenic ED

For men with psychogenic ED, counseling may alleviate the problem. Often, the man and his partner will attend counseling sessions, especially if there is discord in the relationship between the man and his partner. Psychotherapy with a counselor is often very effective for men who have performance anxiety as a cause of their ED.[6] Often, men with psychogenic ED can have their erections jump-started with oral phosphodiesterase-five inhibitor medications (Viagra, Levitra, or Cialis).

Some clues indicate that you may have a psychogenic component to your ED. Men with premature ejaculation and ED may have a psychogenic element to their ED (see Chapter 5.) Suppose a man has a partner-specific ED with one partner and not with other partners. In that case, it is a strong suggestion of psychogenic ED. Finally, if a man masturbates and achieves a full erection with normal ejaculation, this also suggests psychogenic ED.

One of the most common causes of psychogenic ED is performance anxiety. Performance anxiety usually appears in young men after failing to achieve an erection. Anxiety provokes an increase in adrenaline levels in the blood. This hormone constricts the blood supply to the penis and makes having an erection difficult. The negative emotions of these events are stored in the brain and repeat themselves every time a man with performance anxiety attempts sexual intercourse. Unfortunately, the problem worsens each time the man tries sexual activity, creating a vicious cycle. The man will lose his self-confidence on his next attempt to participate in sexual intimacy.

Once psychogenic ED has been diagnosed, help is necessary to reduce anxiety. Counseling by a professional to overcome performance anxiety is essential. We suggest that if you embark on sex counseling, check the therapist's credentials and ask for references. Perhaps the therapist will even arrange for you to speak to a man who has successfully completed sex therapy. This isn't as difficult as you might think, as men who have been helped are often willing and sometimes eager to speak to another man suffering from psychogenic ED. We believe it is important to look for a therapist who focuses on treating sexual problems. Professionals often hold degrees in marriage and family therapy, social work, theology, psychology, or medicine and try to treat sexual problems. Still, you want to be in the hands of a certified sex therapist who has experience treating psychogenic ED.

What Happened to Mr. Johnson?

Mr. Johnson went to his primary care physician, who arranged for him to see a nutritionist and started him on a diet program.

He lost thirty-five pounds in four months. He also started an exercise program with a trainer. As a result of the weight loss and exercise program, his glucose and hemoglobin A1C became normal. His cholesterol and blood pressure also returned to normal. He was provided with a prescription for Viagra, fifty milligrams. He had a response after using the medication on the second try. He found that he didn't need the medication after the weight loss. He reports that his marriage is back on track. He has more energy, an enhanced libido, and improved sexual performance. Both Mr. and Mrs. Johnson are living happily ever after!

Bottom Line

Erectile dysfunction is a common condition affecting millions of American men. The diagnosis is easily made. Nearly every man with this problem can be helped. Today, no one needs to suffer the "tragedy of the bedroom." Call your doctor for an evaluation and treatment; you will be much happier, and so will your partner!

References

1. Feldman, Henry A., Irwin Goldstein, Dimitrios G. Hatzichristou, Robert J. Krane, and John B. McKinlay. "Impotence and Its Medical and Psychosocial Correlates: Results of the Massachusetts Male Aging Study." *Journal of Urology* 151, no. 1 (January 1994): 54-61. https://doi.org/10.1016/S0022-5347(17)34871-1.
2. Schultheiss, Dirk, ed. *Classical Writings on Erectile Dysfunction: An Annotated Collection of Original Texts from Three Millennia.* Berlin: ABW Wissenschftsverlag, 2007.
3. Medical News Today. "Does Marijuana Cause Erectile Dysfunction?" Reviewed August 9, 2018. https://www.medical newstoday.com/articles/31704.

4. Gruenwald, Ilan, Boaz Appel, and Yoram Vardi. "Low-Intensity Extracorporeal Shock Wave Therapy—A Novel Effective Treatment for Erectile Dysfunction in Severe ED Patients Who Respond Poorly to PDE5 Inhibitor Therapy." *Journal of Sexual Medicine* 9, no. 1 (January 2012): 259–64. https://doi.org/10.1111/j.1743-6109.2011.02498.x.

5. Adeldaeim, Hussein M., Tamer Abouyoussif, Omar El Gebaly, Akram Assem, Moataza M. Abdel Wahab, Hazem Rashad, Mostafa Sakr, and Abdel Rahman Zahran. "Prognostic Indicators for Successful Low-Intensity Extracorporeal Shock Wave Therapy Treatment of Erectile Dysfunction." *Urology* 149 (March 2021): 133–39. https://doi.org/10.1016/j.urology.2020.12.019.

6. Atallah, Sandrine, Asad Haydar, Teddy Jabbour, Peter Kfoury, and Georgio Sader. "The Effectiveness of Psychological Interventions Alone, or in Combination with Phosphodiesterase-5 Inhibitors, for the Treatment of Erectile Dysfunction: A Systematic Review. *Arab Journal of Urology* 19, no. 3 (May 2021): 310–22. https://doi.org/10.1080/2090598X.2021.1926763.

CHAPTER 5

Ejaculatory Dysfunction

Bobby is a thirty-five-year-old man with a long history of rapid ejaculation. It is a source of great embarrassment and causes friction between him and his wife. They are also interested in having children, and his wife believes that the problem with his ejaculation is contributing to their problem of a barren marriage.

Bobby is not alone. Approximately 30 percent of American men over age eighteen suffer from premature ejaculation.[1] Problems with ejaculation are more common than erectile dysfunction (or impotence.) It is a problem that is associated with great embarrassment and a contributor to disharmony between partners. This chapter will define premature ejaculation, discuss the causes, and offer treatment suggestions that can resolve this common problem that affects so many men. We will also discuss delayed (impaired) ejaculation and retrograde ejaculation, backwards flow of fluid into the bladder during ejaculation.

Sexual dysfunction pertains to more than erection problems and impotence. Let's analyze how ejaculation occurs, the most common ejaculatory problems, and treatments available for ejaculatory dysfunction (EjD).

How does ejaculation work?

During normal ejaculation, the fluid goes forward through the urethra, the tube in the penis that transports both semen and urine to the outside of the body. At the same time, internal sphincter muscles close off the opening of the bladder to prevent semen from entering the bladder. This function of the internal muscles is necessary so that the seminal fluid moves outside the body, and does not go backward in a retrograde fashion or mix with urine, making the possibility of conception and pregnancy impossible.

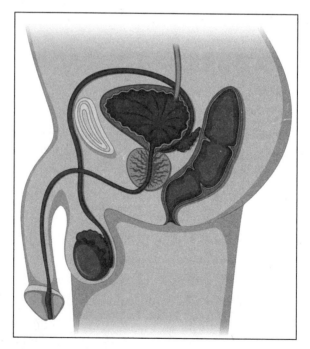

Figure 1: Male Pelvic Anatomy *(Shutterstock)*

When you're aroused, tubes called the vas deferens contract, and the sperm in the tubes from the testes move the sperm toward the back of the urethra. At the same time, the seminal vesicles also release their fluids to mix with the sperm (Figure 1). The urethra senses the sperm and fluid mixture. Then, at the height of sexual excitement, it sends signals to your spinal cord, which in turn sends signals to the muscles at the base of your penis. These muscles contract powerfully and quickly, every eight-tenths of a second. This forces the semen out of the penis as a man reaches his climax.[2]

What is male orgasm?

An orgasm consists of three steps: arousal, orgasm, and resolution. Perhaps the largest sex organ in a man is not between his legs but between his ears, i.e., the brain. When a man becomes sexually aroused, the brain sends an electrical message down the spinal cord to the nerves that supply the penis, prostate, and testes. With arousal there is opening or dilating of the blood vessels to the penis. During this process, more blood rushes into the penis than leaves the penis, and the penis becomes erect. There is also a contraction of the muscles of the scrotum, which pulls the scrotum towards the body, and the muscles throughout the body increase in tension.

This arousal stage lasts from a few seconds to a few minutes. Prior to ejaculation, a clear fluid is deposited into the urethra or the tube in the penis that transports semen and urine from inside the body to the outside of the body. This pre-ejaculatory fluid is meant to improve the chances of sperm survival.

The orgasm itself occurs in two phases: emission and ejaculation. In emission, the man reaches ejaculatory inevitability, often called the "point of no return." Think of ejaculatory inevitability as that point in which you are in bed with your partner and there's a knock at the door; the President of the United States wants to meet with you. You are so involved in sexual intimacy that you don't hear anything. If you hear the knock, you will likely continue in sexual intimacy, ignoring your Commander-in Chief.

Once the spinal reflex is initiated, ejaculation cannot be aborted. The seminal fluid contains prostate fluid, fluid from the ejaculatory ducts, and sperm. It is all deposited near the top of the urethra, ready for ejaculation. Ejaculation occurs in a series of contractions of the penile muscles and around the base of the anus. With the release of fluid there is the transmission of a message back to the brain that is interpreted as heightened enjoyment or pleasure. This is the male orgasm.

After ejaculation, the penis begins to lose its erection. About half of the erection is lost immediately, and the rest fades soon after several more seconds. Muscle tension diminishes, and the man may feel relaxed or drowsy. Men usually must undergo a refractory period, or recovery phase, during which they cannot achieve another erection. This usually lasts for about thirty minutes but may be much longer as a man ages.[3]

Ejaculation involves coordinated muscular and neurological events that involve deposition of semen in the urine channel (emission) and ejection of the fluid from the urethral meatus (ejaculation proper). Emission is accomplished by contraction of the vas deferens, seminal vesicles, and ejaculatory ducts. This process is under adrenaline control. Ejaculation results from the rhythmic contractions of the muscles around the urethra, which causes the forcible ejection of the ejaculate. Within the spinal cord lies the ejaculation center, the area involved in the coordination of signals to and from the brain and penis that eventually lead to ejaculation.

The full sexual response cycle consists of arousal, plateau, orgasm, and resolution and is shown in Figure 2.

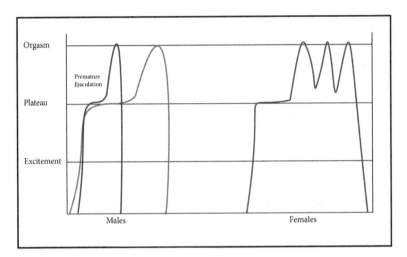

Figure 2: Male Sexual Response Cycle

Notice that there is a distinct time difference between male and female sexual response; women require more time to be aroused than do men. Also note that women can be multi-orgasmic, and this rarely occurs in men. The sexual response cycle for men with premature ejaculation is shown in Figure 2. Their time to reach orgasm is shorter than normal, with minimal or no time in the plateau phase of the sexual response cycle.

Ejaculatory Dysfunctions

There are four main types of ejaculation: 1) premature ejaculation; 2) retrograde ejaculation; (3) impaired ejaculation (orgasm); (4) failure of ejaculation (anejaculation).

Premature Ejaculation

The time from vaginal penetration to ejaculation in the average American male is nine minutes. Premature ejaculation, also known as rapid ejaculation, lacks a definition that is agreed upon by all physicians, but it essentially means the condition whereby a patient ejaculates with minimal sexual stimulation and before he wishes it to occur. Another working definition of premature ejaculation is reaching orgasm in one minute or less after vaginal penetration (intercourse), or ejaculation that occurs too early for partner satisfaction.

This problem is common, occurring in thirty percent of adult men at some time in their lives, and it is the most common form of male sexual dysfunction. It can be caused by erectile dysfunction, anxiety, and nerve hypersensitivity and is generally quite treatable. There are numerous theories as to the dominant cause, but most cases are probably multi-factorial with a contribution from both psychological and physical factors. The management of this problem is best handled with a combination of psychotherapy and medication.

The phenomenon of premature ejaculations is probably as old as humanity. Writings as early as Greek antiquity made mention of an *ejaculatio ante portas*. But it was not until the late nineteenth century that the experience was described in the medical literature and understood as a medical disorder or a medical condition. Premature ejaculation has also been called early ejaculation, rapid ejaculation, rapid climax, premature climax, and (historically) ejaculation praecox.

Premature ejaculation is a common sexual complaint. Estimates vary, but as many as one out of three men say they have experienced this problem at some time during their lives. Premature ejaculation is the most common sexual problem occurring in men under the age of forty. As long as it happens infrequently, it's not cause for concern. However, when it occurs all or the majority of the time, it is a source of anxiety and tension, and it can contribute to discord between a man and his partner.

A man may meet the diagnostic criteria for premature ejaculation if he meets these thresholds:

- Always or nearly always he ejaculates within one minute of penetration.
- He is unable to delay ejaculation during intercourse all or nearly all of the time.
- He feels distressed and frustrated and tends to avoid sexual intimacy as a result.

There are two classifications of premature ejaculation: lifelong (primary) or acquired (secondary). Lifelong premature ejaculation occurs all or nearly all of the time, beginning with your first sexual experience. Acquired premature ejaculation has the same symptoms but develops after you've had previous sexual experiences without ejaculatory problems.

Not all men can fit into either of these categories. Many men feel that they have symptoms of premature ejaculation, but the symptoms do not meet the diagnostic criteria for premature ejaculation. Instead, they may have variable premature ejaculation, which is characterized by periods of rapid ejaculation as well as periods of normal ejaculation. For example, a man may have premature ejaculation with one partner and not with another partner, nonetheless still causing him great alarm and concern.

Causes of Premature Ejaculation

The exact cause of premature ejaculation isn't known. While it was once thought to be only psychological in origin, doctors now know that premature ejaculation is more complicated and involves a complex interaction of psychological and biological factors.

Psychological Causes

Early sexual experiences may establish a pattern that can be difficult to change later in life. For example, there are situations in which you may have hurried to reach climax in order to avoid being discovered in the act of sexual intimacy. This is common among young men with their first few sexual encounters. The rapid ejaculation becomes a pattern and the brain is trained to complete the act quickly. Later in life, this pattern is difficult to alter and premature ejaculation continues.

Another example is a man who has received advice that sexual intercourse should not take place until after marriage. If the man participates in sexual intimacy, there may be guilt associated with intercourse that may increase his tendency to rush through sexual encounters.

Other factors that can play a role in causing premature ejaculation include:

- Erectile dysfunction: Men who are anxious about obtaining or maintaining an erection during sexual intercourse may form a pattern of rushing to ejaculate, which can be difficult to change. Also, when a man struggles to stay stimulated enough to maintain his erection, he may hasten the time to reach climax.

- Anxiety: Many men with premature ejaculation also have problems with anxiety, either specifically about sexual performance or related to other issues.
- Relationship problems: If you have had satisfying sexual relationships with other partners in which premature ejaculation happened infrequently or not at all, it's possible that interpersonal issues between you and your current partner are contributing to the problem.

Biological Causes

While psychological factors are the most common causes, a number of biological factors may also contribute to premature ejaculation.

The most common biological cause is an infection of the prostate gland and/or the urethra. Inflammation of the prostate gland can make the prostate more sensitive and trigger a rapid ejaculation. Abnormal levels of brain chemicals called neurotransmitters such as serotonin may be associated with premature ejaculation. Also, too much thyroid hormone or hyperthyroidism may contribute to premature ejaculation.

Other causes of PE related to certain medications and also associated with the use of alcohol and recreational drugs need to be evaluated.[4]

There is also a possibility that premature ejaculation can be an inherited problem, but it is difficult for most men to obtain information from their father or grandfather about their sexual experiences. Finally, nerve damage from surgery or trauma is a rare cause of premature ejaculation.

We have listed the most common causes of premature ejaculation. However, there are other factors that can increase your risk of premature ejaculation.

You may be at increased risk of premature ejaculation if you occasionally or consistently have trouble getting or maintaining an erection. Fear of losing your erection may cause you to consciously or unconsciously hurry through sexual encounters. (For more information on erectile dysfunction, see Chapter 4.)

If you have a serious or chronic medical condition, such as heart disease, you may feel anxious during sex and may unknowingly rush to ejaculate. Often a discussion with your doctor about engaging in sexual intimacy can alleviate those fears and concerns about engaging in sexual intimacy.

Emotional or mental strain in any area of your life can play a role in premature ejaculation, often limiting your ability to relax and focus during sexual encounters. Engaging in sexual activity requires a certain level of relaxation to make the nervous system function properly. Stress can be a distraction that impairs the proper function of the nerves that supply the penis and the prostate gland, creating a situation that results in premature ejaculation.

Other Manifestations of Premature Ejaculation

While premature ejaculation alone doesn't increase your risk of health problems, it can cause significant problems in your personal life and can be a source of personal distress not only to you but also to your partner. Men who are not in a relationship appear to have the greatest distress, as the condition often precludes the man from seeking out or initially maintaining a relationship.

Partners often complain that the man with premature ejaculation is focusing on his performance and not on his partner. Significant numbers of partners report that the man's premature ejaculation resulted in a failure of the relationship.[5]

Probably the most important and common side effects of premature ejaculation are embarrassment, anxiety, stress in the relationship, and even depression.[6]

Premature ejaculation can occasionally make fertilization difficult or impossible for couples that are trying to achieve a pregnancy. Bobby's history is not unusual; premature ejaculation can result in failure of the sperm to be properly deposited in the vagina and thus prevent the sperm from reaching the woman's egg and allowing natural fertilization to take place.

The Evaluation of Premature Ejaculation

In addition to asking about your sex life, your doctor will ask about your health history and may perform a general physical exam. The health history will consist of questions similar to those shown in Table 1. Your doctor may order a urine test

---

Table 1
Questions you might be asked by your doctor evaluating premature ejaculation:

1. What is the estimated time from penetration to ejaculation?

2. When was your first sexual experience?

3. Do you have an adequate erection which enables to make penetration?

4. Is the problem specific to one partner specific?

5. Do you have problem with rapid ejaculation during masturbation?

6. What is the impact of your problem on your partner?

7. What previous treatment(s) have you had?

8. Is the problem causing anxiety, embarrassment, depression?

9. Is the problem associated with any other medical conditions?

---

to rule out possible infection and a prostate exam to see if the prostate gland is the culprit. If you have both premature ejaculation and trouble getting or maintaining an erection, your doctor may order blood tests to check your male hormone (testosterone) levels or other blood tests to test, for example, your thyroid level.

Treatment of Premature Ejaculation

Common treatment options for premature ejaculation include behavioral techniques, topical anesthetics, oral medications, and counseling. One or more treatment options are often required to treat premature ejaculation.

DIY Solutions for Premature Ejaculation

In some cases, therapy for premature ejaculation may involve taking simple steps such as masturbating an hour or two before intercourse so that you're able to delay ejaculation during sex. Your doctor also may recommend avoiding intercourse for a period of time and focusing on other types of sexual play so that pressure to perform sexually is removed from your sexual encounters.

The Pause-Squeeze Technique

Your doctor may instruct you and your partner in the use of a method called the pause-squeeze technique. This technique, which was developed by Masters and Johnson, pioneering sex therapists in the 1960s, pushes blood out of the penis and momentarily decreases sexual tension and represses the ejaculatory response. It has been successful in some men with premature ejaculation.[7]

The pause-squeeze technique is easy to learn. You begin sexual activity as usual, including stimulation of the penis by yourself or by your partner, until you feel almost ready to ejaculate. When you feel close to having an ejaculation, have your partner squeeze the end of your penis, at the point where the head (glans) joins the shaft, and maintain the squeeze for several seconds, until the urge to ejaculate passes. Usually, a five to seven second squeeze is adequate. After the squeeze is released, wait for about thirty seconds, and then resume foreplay. You may notice that squeezing the penis causes it to become less erect, but when sexual stimulation is resumed, it soon regains a full erection.

By repeating this as many times as necessary, you can reach the point of entering your partner without ejaculating. After a few practice sessions, the feeling of knowing how to delay ejaculation may become a habit that no longer requires the pause-squeeze technique.

Masturbation before Intimacy

This is a time-tested treatment that is effective as well as inexpensive. One caveat: as a man ages, the time from one ejaculation to the next erection, referred to as the refractory period, increases in duration. That means you'll need to allow yourself more time between masturbation and sexual intimacy than you did back in your teen years. Failure to follow this advice may result in trading your premature ejaculation for impotence, which is no less embarrassing or discouraging. How do you find out? You have to experiment and find that precious time interval that works from time of masturbation to intercourse.

Try not to sweat the small stuff. Just putting pressure on yourself may be the culprit causing your PE. The likelihood of having simultaneous orgasms is as rare as hen's teeth. It just doesn't happen the majority of the time. You can allay lots of apprehension if you just relax and enjoy the moment of sexual intimacy without the pressure of having an orgasm at the same time as your partner.

Distract yourself and don't think about your orgasm. If you notice yourself getting too excited, turn your thoughts to something distant, abstract and unsexy, such as counting sheep, subtracting seven from one hundred down to zero, or the anticipated scores of your favorite football team. Avoid thinking of a topic that is going to make you stressed, like your quota in business or your upcoming bonus, as that may cause you to lose your erection.

Controlling the Orgasm

Orgasm control is the practice of maintaining a high level of sexual arousal while delaying ejaculation. It takes practice, but it gets easier over time. Here are two methods recommended by the National Institutes of Health to stop premature ejaculation:

Stop-and-start method. Have intercourse as usual until you feel yourself coming uncomfortably close to orgasm. Immediately and abruptly cease all stimulation for thirty seconds, then start again. Repeat this pattern until you're ready to ejaculate.

Change it up. Some intercourse positions put less pressure on the glans. Try "passive" positions. Go from missionary or male superior to male on the bottom. Missionary and rear-entry positions place the most stimulation and friction on the glans, so consider taking them off the table until more control is achieved.

Slow down. Depending on your personal sensitivity, slowing your movements and pelvic thrusting and opting for gentler intercourse can help you hold off orgasm. If you find yourself getting too close to orgasm, put the brakes on, change to a new position, or take a pause to stimulate your partner in other ways.

Focus on foreplay. Never forget that you can give your partner a great sexual experience through more extended, intimate, attentive, and generous foreplay. Stimulate your partner enough manually, orally, or with toys such as vibrators, and they may not need a prolonged period to achieve sexual satisfaction.

The Helpful Kegel

Help may be just a Kegel away. Kegel exercises have been recommended to women for decades to control incontinence (the urge to urinate all the time), and to enhance sexual responses during intercourse. Now the Kegel exercise has been demonstrated to be effective for men. Kegel exercises are used to strengthen the muscles responsible for bladder control. These pelvic floor muscles are the same muscles you use to prevent urination, or to keep from passing gas from your rectum at an inappropriate moment.

There's no better method to strengthen the pelvic region in both men and women than to create a strong pubococcygeus muscle (PC muscle), which can help control ejaculation. The easiest way for a man to find this muscle is to see if he can stop the flow of urine when going to the bathroom. It's the PC muscle that you use to do that. Once you find it, you need to practice feeling exactly where it is located and make sure you can contract and relax your PC muscle, rather than using your abdominals, buttocks, or thighs. (These must all stay loose when doing Kegel exercises.) To do Kegels, you will clench and release the PC muscle repeatedly for a slow count to five each time and then relax for ten seconds. If you do the Kegel exercise ten times, that is one set. Perform one set of ten reps, three times per day (thirty total per day). The Kegel exercise can then be performed during sexual intimacy, allowing your PC muscle to prevent orgasm.

After two to three weeks of exercising, you should see improvement. Remember, Arnold Schwarzenegger did not become a Mr. Olympia champion after his first visit to the weight room. The same applies to the man with premature ejaculation. One day of exercises is not going to result in significant improvement in building up the PC muscle. The reality is that it may take several weeks, or even months, to see improvement in delaying ejaculation. If you have PE, it's okay to use your PC! Flexing and strengthening your pubococcygeus (PC) muscle can help you exert more control over ejaculation.

Locate your PC muscle by putting one or two fingers right behind your testicles or your scrotum. Pretend that you are urinating, then try to stop the flow with a quick muscle contraction. That muscle you just used to stop the flow from the bladder is your PC muscle.

Flex that PC muscle regularly. Try to do ten to twenty squeezes in a set, two or three times a day. Do a set whenever you're bored or stationary—as when you're sitting at your desk or in traffic. It's a secret; no one will be able to see that what you're doing is strengthening your PC muscles. Squeeze your PC muscle when you feel ejaculation is about to occur. Once the muscle is strong enough, you should be able to hold it off just like stopping flow when urinating.

Practice controlling your orgasm. Either with a patient partner or when masturbating, focus on improving your control over your climax. Stimulate yourself to the brink of climax and then stop. Do this several times before finishing it off with an orgasm. As you practice, learn to recognize the feeling of getting close to orgasm, and take note of how close you can get and still have control and stop (just prior to reaching that "point of no return"). During intercourse, use that knowledge to slow down or adjust your movements if you get too close too early.

Condoms are not just for prevention of sexual transmitted diseases (STDs). Something as simple as an over-the-counter or machine-generated condom works for a lot of men with PE. Condoms decrease the stimulation of the very sensitive glans, or head of the penis, for most men, which will often prolong the time before ejaculation.

Some condoms are coated with a slight numbing gel on the inside. This can help you prolong ejaculation without causing numbness to your partner. If you opt for this technique, just make sure you know which side is where when you put it on because the wrong side puts your partner on the wrong side of the orgasm track.

Topical Anesthetics

Anesthetic creams and sprays that contain a numbing agent, such as lidocaine or prilocaine, are often used to treat premature ejaculation. These products, which can be obtained over the counter and through the internet, are applied to the penis a short time before sexual intimacy to reduce sensation and thus help delay ejaculation. Unfortunately, some of these preparations may be transferred to the partner, thus making sexual enjoyment for the partner diminished or absent because of numbness of the vagina and particularly the clitoris. The only solution is to apply the topical anesthetic and then put on a condom to avoid transfer to the partner.

A lidocaine spray for premature ejaculation (Promescent) is now available over the counter. Promescent is a topical medication that is applied to the penis ten minutes before sexual activity, and it helps a man to better manage the sensations of sex through desensitization. However, unlike other topical medications for early ejaculation, Promescent penetrates only the most superficial layer of skin on the head of the penis, or the glans, and is not absorbed through the skin on the shaft of the penis. Since it is not absorbed into the skin, it will not negatively impact your partner's sensations or result in decreased sensation for the partner.

Severance Secret cream (SS-cream) is formulated from nine natural substances including ginseng and cinnamon, and it has local desensitizing effects. Studies have demonstrated that the effect persists for up to two hours. For best results the cream is applied one hour before engaging in sexual intimacy. The side effects of SS-cream are mild local burning and mild pain at the head of the penis. There are no adverse effects on sexual function, or any loss of sensation reported by the man's partner.[8]

Although topical anesthetic agents are effective and well-tolerated, they have potential side effects. Some men report temporary loss of sensitivity and decreased sexual pleasure. In some cases, female partners also have reported these effects of lack of sensation. In rare cases, lidocaine or prilocaine can cause an allergic reaction.

Oral Medications

Many oral medications may delay orgasm. Although none of these drugs is specifically approved by the Food and Drug Administration to treat premature ejaculation, some are used for this purpose, including antidepressants, analgesics, and phosphodiesterase-5 inhibitors which include Viagra, Cialis, and Levitra. These medications may be prescribed for either use thirty to forty-five minutes before sexual intimacy or daily use, and may be prescribed alone or in combination with other treatments.

Many years ago, it was noted that one of the side effects of certain antidepressants is delayed orgasm. For this reason, selective serotonin reuptake inhibitors (SSRIs), such as sertraline (Zoloft), paroxetine (Paxil) or fluoxetine (Prozac, Sarafem), are used to help delay ejaculation. If SSRIs don't improve the timing of your ejaculation, your doctor may prescribe the tricyclic antidepressant clomipramine (Anafranil). The dosage of these medications used for treating premature ejaculation is shown in Table 2.

Table 2
SSRI medications used to treat premature ejaculation.

Drug (Trade name)	Dosage before intercourse	Side effects
Clomipramine (Anafranil)	25-50mg	nausea
Paroxetine (Paxil)	20mg	sleepiness, yawing
Sertraline (Zoloft)	50mg	headache, fatigue
Fluoxetine (Prozac)	5-20mg	headache, insomnia

Table 2: Dosages of SSRIs for Treating PE

These drugs can be taken several hours before sexual activity, and since they inhibit arousal, they can help make it easier for a man to control ejaculation. Before he makes a decision regarding these drugs that require a prescription, he will need to see either his general practitioner or a urologist; he should also ask about the side effects of SSRIs. Unwanted side effects of antidepressants may include nausea, dry mouth, drowsiness, and decreased libido. Also, these medications should not be abruptly discontinued.

The Possibility of Priligy

Dapoxetine (Priligy) is the only drug that has been created specifically for the treatment of premature ejaculation by increasing the neurotransmitter serotonin, which can be effective in prolonging the time from penetration to ejaculation. The drug has been approved in more than fifty countries for the treatment of premature ejaculation but has not received approval from the FDA at the time of publication of this book. The tablets, 30mg to 60mg, taken one to two hours before sexual intimacy, have shown to be effective, even at the first dose, with an improvement of the time from penetration to ejaculation. The drug has been reported to be effective in men with life-long premature ejaculation as well as men who acquired premature ejaculation later in life. Also, the drug has been effective in men with premature ejaculation and erectile dysfunction treated with any of the phosphodiesterase-5-inhibitors (Viagra, Levitra, Cialis). The side effects occur in less than 3 percent of the men using Dapoxetine and consist of transient nausea, dizziness, and diarrhea.

Tramadol (Ultram) is a medication commonly used to treat pain. It also has side effects that delay ejaculation. Tramadol may be prescribed when SSRIs haven't been effective. Unwanted side effects of tramadol may include nausea, headache, and dizziness. How tramadol works has not been completely elucidated. Perhaps the mechanism of action is the anesthetic-like effect. Because of the potential of tramadol to cause addiction, its off-label usage has been discouraged. However, tramadol may be considered when other treatments have failed.

Using ED Drugs to Treat Premature Ejaculation

Some medications used to treat erectile dysfunction, such as sildenafil (Viagra, Revatio), tadalafil (Cialis, Adcirca), or

vardenafil (Levitra, Staxyn), also may help premature ejaculation. By enhancing erectile function, less focus can be placed
on maintaining an erection, thereby reducing the possibility
of overstimulation. Unwanted side effects may include headache, facial flushing, temporary visual changes, and nasal
congestion.[9]

These oral medications can be used alone for treating ED
or in combination with the SSRIs as a treatment for premature
ejaculation. These ED drugs are particularly effective in men
who have both ED and premature ejaculation. It is worth noting that nearly fifty percent of men with ED also have a problem with premature ejaculation.[10]

Alpha-blockers to Unblock Premature Ejaculation

Older or middle-aged men with urinary symptoms such as frequency of urination, getting up at night to urinate, and dribbling
after urination as a result of the enlarged prostate gland who also
have premature ejaculation, may be effectively treated with drugs
that block the nerves to the prostate gland muscles that control
ejaculation. This is a nice example of a "twofer"; one drug used to
treat two conditions. However, these drugs can also cause retrograde ejaculation as a side effect (discussed later in this chapter.)

Drugs such as tamsulosin (Flomax), siladosin (Rapaflo) or
alfuzosin (Uroxatrol) are alpha-blockers and have been used for
decades for treating the enlarged prostate gland (see Chapter
2 on prostate gland enlargement). These drugs effectively
block the nerve transmission of nerve impulses to the prostate
gland, not only decreasing the resistance to the flow of urine
from the bladder to the outside of the body but also decreasing the sensitivity to the prostate gland. Ejaculation is a reflex
that may be impacted using these alpha-blocker medications.

Documentation indicates that alpha-blockers may be effective in delaying ejaculation in approximately 50 percent of the men using these alpha-blocking drugs.[11]

Psychological Interventions

This approach, also known as talk therapy, involves talking with a mental health provider, such as a social worker, psychologist, or sex therapist about your relationships and experiences. These sessions can help you reduce performance anxiety and find better ways of coping with stress. Counseling is most likely to help when it's used in combination with drug therapy. Couples can attend the sessions together, or the counselor may suggest that the man and partner are seen separately.

Counseling with a therapist helps men develop the skills that help them to delay ejaculations while increasing the man's self-confidence and decreasing performance anxiety. Counseling can also help to resolve interpersonal issues that may have precipitated the premature ejaculation or have occurred as a result of the sexual problem. Consultations may occur one-on-one with the therapist, with both the man and his partner being treated simultaneously, or in a group environment with several couples being treated at the same time by the therapist. It has been our observation that a man who is not in a relationship is more difficult to treat than a man who is in a relationship and has a sexual partner. It is also our experience that premature ejaculation is more effectively treated in conjunction with medical therapy than with counseling alone.

Alternative Treatments for Premature Ejaculation

Several alternative medicine treatments have been studied, including selected behavioral therapies (yoga). Additionally, surgical treatments such as surgically dividing the nerves to the head of the penis to decrease its sensitivity have been done, but only on men who have been resistant to conservative treatment with medication and/or counseling. The role of surgery in the management of premature ejaculation remains unproven; until the results of further studies have been reported, we do not recommend this approach to treatment.

Yoga is a popular form of complementary and alternative treatment. Use of yoga for various bodily ailments is recommended in ancient *Ayurveda* (*ayur*=life, *veda*=knowledge) texts and is being increasingly investigated scientifically. Many patients and yoga protagonists claim that it is useful in men with premature ejaculation. Several studies have documented that yoga appears to be a feasible, safe, effective, and acceptable non-pharmacological option for premature ejaculation.

Herbal Remedies and Supplements

Some herbal remedies may help to alleviate the problem. Take one hundred milligrams of kava before engaging in sexual intercourse. Kava root comes in pill form and powder form, which can be brewed into a tea; it is sold at many health food stores. Kava increases blood flow to the penis and slows reaction to increased sexual stimulation, therefore helping you maintain an erection longer.

Add two drops of hibiscus flower essence to one-quarter cup of water and sip slowly. Hibiscus flower essence helps relieve stress and promotes romantic sexual feelings, which increases the ability to maintain erection and control ejaculation.

Take a daily multivitamin formulated for men. Using multivitamin supplements in conjunction with a healthy diet and exercise may increase sexual stamina and performance.

Take Graphites 12C two to four times daily. This nutrient in supplement form will help stop premature ejaculation with repeated use. Once your condition has improved, you may discontinue use of this supplement.

Try taking a 5HTP (hydroxytryptophan) supplement daily. 5HTP is a naturally occurring nutrient that is responsible for sexual health and serotonin production. It suppresses the urge to ejaculate sooner than normal, and it increases stamina. You can find it at many health food stores and pharmacies.

Retrograde ejaculation

Retrograde ejaculation is the process whereby the semen is passed in a retrograde or backward direction into the bladder rather than exiting the urethra. There are three potential causes to this problem: anatomic (following surgery on the prostate gland or from a congenital process), neurologic (due to disorders that interfere with the ability of the bladder neck to close during emission such as diabetes mellitus, or surgery that can affect the nerves to the bladder and prostate gland) and side effects of certain medications that cause paralysis of the internal muscles or sphincters. This condition is diagnosed by the finding of seminal fluid and/or sperm in a urine specimen obtained immediately after orgasm.

The treatment of retrograde ejaculation depends to some extent on the cause. Anatomic causes are rarely curable, and sperm harvesting from the bladder is required for those men with retrograde ejaculation wishing to initiate a pregnancy. Pharmacologic causes are generally reversible by withdrawal of the offending medication. Neurologic causes are difficult to treat if there is complete nerve damage, such as that which may occur in spinal cord injuries. In those patients with a partial neural injury (diabetes), the use of certain medications (ephedrine and pseudoephedrine) may convert the man with retrograde ejaculator to an antegrade, or normal, ejaculator.

Delayed Ejaculation (DE)

A common problem in older men is delayed or impaired ejaculation where it takes more than twenty-five minutes to achieve an orgasm. This can be a source of tension and pressure between a man and his partner. Your partner might feel that you're not attracted sexually and that love has gone out the bedroom door. You might feel frustrated or embarrassed about wanting to achieve ejaculation but being physically or mentally unable to do so. Treatment or counseling can help resolve these issues.

The causes of delayed ejaculation include conditions and reactions to medications. Psychological causes of DE can occur due to a traumatic experience. Cultural or religious taboos can give sex a negative connotation. Anxiety and depression can both suppress sexual desire, which may result in DE as well.

A myth that has been widely circulated is that circumcision or removing the foreskin of the penis decreases the sensation and can result in delayed ejaculation. Numerous studies have shown that a circumcision does not have adverse effects on sexual function, including DE.[12]

Relationship stress, poor communication, and anger can make DE worse. Disappointment in sexual realities with a partner compared to sexual fantasies can also result in DE. Often, men with this problem can ejaculate during masturbation but not during stimulation with a partner.

Certain chemicals can affect the nerves involved in ejaculation. This can affect ejaculation with and without a partner. These medications can cause DE: antidepressants, such as fluoxetine (Prozac), antipsychotics, such as thioridazine (Mellaril), and medications for high blood pressure, such as propranolol (Inderal). Diuretics and alcohol can also cause DE.

Surgeries or trauma may also cause DE, including the following events: damage to the nerves in your spine or pelvis, certain prostate surgeries like prostate gland removal for cancer that cause nerve damage, heart disease that affects blood pressure to the pelvic region, neuropathy or stroke, low thyroid hormone, and low testosterone levels. A temporary ejaculation problem can cause anxiety and depression. This can lead to recurrence, even when the underlying physical cause has been resolved. Also in rare cases, delayed ejaculation is a sign of worsening health problems such as heart disease or diabetes.

The same tests for premature ejaculation are also indicated for delayed ejaculation. Another simple test is to check the reaction of your penis to a vibrator, which may reveal a physical problem. An easy method of performing this test is to use a tuning fork and check your appreciation of the vibration on your hand compared to the shaft of the penis. If the there is significant decrease in perception of the vibration on the penis compared to the hand, then there is likely a physical explanation of the delayed ejaculation.

At the present time, no drug has been specifically approved for delayed ejaculation, but medications used for conditions such as Parkinson's disease have been shown to help.

Off-label drug use means that a drug that's been approved by the FDA for one purpose is used for a different purpose that has not been approved. Your doctor can still use the drug for that purpose. This is because the FDA regulates the testing and approval of drugs, but not how doctors use drugs to treat their patients. So, doctors can prescribe a drug however they think is best for you.

Some medications have been used to help DE, but none have been specifically approved for it. Several of these medications include cyproheptadine (Periactin), which is an allergy medication, amantadine (Symmetrel), which is a drug used to treat Parkinson's disease, and buspirone (Buspar), which is an antianxiety medication. Of course, treating illicit drug use and alcoholism, if applicable, can also help DE.

Psychological counseling can help treat depression, anxiety, and fears that trigger or perpetuate delayed ejaculation. Sex therapy may also be useful in addressing the underlying cause of sexual dysfunction. This type of therapy may be completed alone or with your partner.

DE can generally be resolved by treating the mental or physical causes. Identifying and seeking treatment for DE sometimes exposes an underlying medical condition. Once this is treated, DE often resolves. The same is true when the underlying cause is a medication. However, don't stop taking any medication without your doctor's recommendation.

Take-home message on delayed ejaculation: Delayed ejaculation does not pose any serious risks to your life; it can, however, be a source of stress and may create problems in your sex life and personal relationships. For most men, help is available.

Whatever Happened to Bobby?

Fortunately, Bobby's wife is very patient and understanding. Their success began simply with conversation. They decided to seek help from a therapist who shared several practicing techniques, along with judicious use of a numbing agent. With several months of practice, Bobby developed the new habit of a more prolonged and satisfying duration of intercourse prior to reaching climax. Their biggest problem will be finding the time for his sexual stamina once the first baby arrives. By the way, they are expecting a new baby!

Bottom Line

Problems with ejaculation are one of the most common ailments to afflict men of all ages. It can wreak havoc on a man's life and certainly impact his partner as well. Help is available, and most men can be helped with both medical and nonmedical management. If you are suffering from ejaculation problems, see your doctor. Premature ejaculation doesn't mean you always have to say, "I'm so sorry."

References

1. Barnes Tricia, and Ian Eardley. "Premature Ejaculation: The Scope of the Problem." *Journal of Sex and Marital Therapy* 33, no. 3 (March-April 2007): 151–70. https://doi.org /10.1080/00926230601098472.
2. WebMD. "How the Penis Works: Erection and Ejaculation." Reviewed May 2, 2023. http://www.webmd.com/erectile-dysfunction /how-an-erection-occurs.
3. Alkon, Cheryl. "Male Orgasm: Understanding the Male Climax." Everyday Health, reviewed August 1, 2022. http://www.every dayhealth.com/sexual-health/the-male-orgasm.aspx.
4. Pugh, J., and S. Belenko. "Alcohol, Drugs and Sexual Function: A Review." *Journal of Psychoactive Drugs* 33, no. 3 (July-September 2001): 223–32. https://doi.org/10.1080/02791072.20 01.10400569.

5. Burri, Andrea, François Giuliano, Chris McMahon, and Hartmut Porst. "Female Partner's Perception of Premature Ejaculation and Its Impact on Relationship Breakups, Relationship Quality, and Sexual Satisfaction." *Journal of Sexual Medicine* 11, no. 9 (September 2014): 2243–55. https://doi.org/10.1111/jsm.12551.

6. Mathew, Roy J., and Maxine L. Weinman. "Sexual Dysfunctions in Depression." *Archives of Sexual Behavior* 11 (1982): 323–28. https://doi.org/10.1007/BF01541593.

7. Masters William H., and Virginia E. Johnson. *Human Sexual Response.* Boston: Little, Brown and Company, 1966.

8. Choi, Hyung K., Gyung Woo Jung, Ki Hak Moon, Zhong Cheng Xin, Young Deuk Choi, Woong Hee Lee, Koon Ho Rha, Yeong Jin Choi, and Dong Ki Kim. "Clinical Study of SS-Cream in Patients with Lifelong Premature Ejaculation. *Urology* 55, no. 2 (February 2000): 257–61. https://doi.org/10.1016 /S0090-4295(99)00415-X.

9. McMahon, Chris G., Bronwyn Stuckey, Morten Andersen, Kenneth Purvis, Nandan Koppiker, Scott Haughie, and Mitra Boolell. Efficacy of Sildenafil Citrate (Viagra) in Men with Premature Ejaculation." *Journal of Sexual Medicine* 2, no. 3 (May 2005): 368–75. https://doi.org/10.1111/j.1743-6109.2005.20351.x.

10. Jannini Emmanuele A, F. Lombardo, and A. Lenzi. "Correlation between Ejaculatory and Erectile Dysfunction. Supplement. *International Journal of Andrology* 28, no. S2 (December 2005): 40–45. https://doi.org/10.1111/j.1365-2605.2005.00593.x.

11. Choi, Jae Hwi, Jung Seog Hwa, Sung Chul Kam, Seong Uk Jeh, and Jae Seog Hyun. "Effects of Tamsulosin on Premature Ejaculation in Men with Benign Prostatic Hyperplasia." *World Journal of Men's Health* 32, no. 2 (August 2014): 99–104. https://doi.org/10.5534/wjmh.2014.32.2.99.

12. Collins, S., J. Upshaw, S. Rutchik, C. Ohannessian, J. Ortenberg, and P. Albertsen. "Effects of Circumcision on Male Sexual Function: Debunking a Myth? *Journal of Urology* 167, no. 5 (May 2002): 2111–12. https://doi.org/10.1016/S0022-5347(05)65097-5.

CHAPTER 6

Considering a Vasectomy

Paco is forty-three years old and has three children; his wife has been on birth control pills for the past eight years, and they believe that their family is complete. He is giving thought to having a vasectomy.

Vasectomy is a minor surgery to block sperm from reaching the semen ejaculated from the penis. Semen will still exist, but it will contain no sperm. In fact, most of the volume of semen originates from the prostate, and two adjacent glands called the seminal vesicles. The testicles contribute only three percent of the total volume of the ejaculate. All the sperm are contained in this three percent. After a vasectomy the testes still make sperm, but the sperm are soaked up by the body. Each year, more than half a million men in the US choose vasectomy for birth control. A vasectomy prevents pregnancy better than any other method of birth control, except abstinence. Only one to two women out of a thousand will get pregnant in the year after their partners have had vasectomies.

What Happens Under Normal Conditions?

Both sperm and male sex hormones are made in the paired testes (testicles). The testes are in the scrotum at the base of the penis. Sperm leave the testes through a coiled tube (the "epididymis"), where they stay until they're ready for use. Each epididymis is linked to the prostate by a long tube called the vas deferens (or "vas"). The vas runs from the lower part of the scrotum into the inguinal canal located in the groin. It then goes into the pelvis and behind the bladder. This is where the vas deferens joins with the seminal vesicle and forms the ejaculatory duct. When you ejaculate, seminal fluid from the seminal vesicles and prostate mix with sperm to form semen. The semen flows through the urethra and exits from the end of your penis.

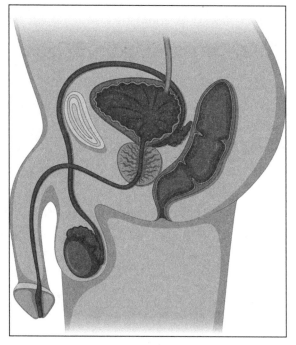

Figure 1: Male Pelvic Anatomy *(Shutterstock)*

Vasectomy consists of dividing the vas (the tube that delivers the sperm from the testis to the prostate) to prevent conception. It is the most common method of male contraception in this country. Since vasectomy simply interrupts the delivery of the sperm, it does not change the hormonal function of the testis. Thus, the sexual drive (libido) and ability to engage in sexual intimacy remain intact and will not be affected by the vasectomy. Since most semen is composed of fluid from the prostate and seminal vesicles, the semen volume will look the same. Vasectomy is thought to be free of known long-term side effects. It is considered the safest and most reliable method of permanent male sterilization.

The no-scalpel vasectomy was developed in 1974 by a Chinese physician, Dr. Li Shunqiang, and it has been performed on over eight million men in China.

After injecting the scrotal skin and each vas with a local anesthetic, we use a special vas-fixation clamp to encircle and firmly secure the vas without penetrating the skin. One blade of a forceps or clamp is used to penetrate the scrotal skin. The tips of the forceps are spread, opening the skin much like spreading apart the weaves of fabric. The opening is approximately one-quarter inch in length. The vas is thus exposed and then lifted out and occluded by standard techniques, such as cautery or sutures. The second vas is then brought through the same opening and occluded similarly. The skin wound contracts to a few millimeters and usually does not require suturing. Patients should feel no discomfort as no needle is used to insert the local anesthetic, and the opening is about the size of the tip of a pencil.

Compared to the traditional incision technique, the no-scalpel, no-needle vasectomy usually takes less time, causes less discomfort, and may have lower bleeding rates and infection. There is minimal pain or discomfort as no needle is used to insert the local anesthetic, and no-scalpel is used to make the opening in the scrotum. Recovery following the procedure is usually complete in two to three days. Hard work or straining (athletic pursuits or heavy lifting) is not recommended for seven days. Most men should refrain from sexual intimacy for a week after the procedure.

Common Reasons for Having a Vasectomy

1. You want to enjoy sex without worrying about pregnancy.
2. You do not want to have more children than you can care for.
3. Your partner has health problems that might make pregnancy difficult.
4. You do not want to risk passing on a hereditary disease or disability.
5. You and your partner don't want to or can't use other kinds of birth control.
6. You want to save your partner from the surgery involved in having her tubes tied, and you want to save the expense.

Common Questions Asked Before Selecting a Vasectomy

"How Can I Be Sure That I Want a Vasectomy?"

You must be absolutely sure that you don't want to father a child in the future. You should talk to your partner. It is good to make this decision together, consider other kinds of birth control, and speak to friends or relatives who may have had a vasectomy. You and your partner might want to think about how you would feel if your partner had an unplanned pregnancy. Talk to your doctor, nurse, or family planning counselor.

A vasectomy might not be suitable for you if you are very young, if your current relationship is not permanent, or if you are having a vasectomy just to please your partner. You do not really want to consider a vasectomy if you are under a lot of stress or are counting on being able to reverse the procedure later.

"How Does My Vasectomy Prevent Pregnancy?"

Sperm is made in the man's testicles. The sperm then travels from the testicle through a tube called the vas, where it enters

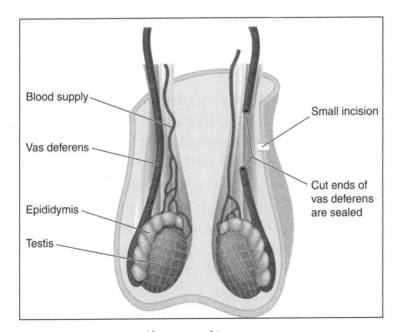

Figure 2: Vasectomy *(Shutterstock)*

the prostate gland. In the prostate, the semen is made, and here the sperm mixes with the semen. The prostate is connected to the penis. The sperm and semen are ejaculated into the urethra and then out the end of the penis. In a vasectomy, the vas, or tube, is blocked so that sperm cannot reach the prostate to mix with the semen. Without sperm in the semen, a man cannot impregnate his partner. Figure 2 shows the vas before and after the vasectomy.

"What Preparation Must I Take Before My Vasectomy?"

We suggest that you wash the scrotum with any soap to prevent infection. You should bring a pair of tight-fitting jockey shorts or an athletic supporter to support the scrotum and minimize swelling after the procedure. We suggest that you avoid using any anti-inflammatory drugs such as aspirin, ibuprofen, and fish oil one week before the procedure. These drugs and supplements may thin the blood and cause excessive bleeding during or after the procedure.

"What is Different about the No-Scalpel Vasectomy?"

The no-scalpel vasectomy is significantly different from a conventional vasectomy. An improved method of anesthesia helps make the procedure less painful. In a conventional vasectomy, the physician makes one or two small incisions approximately one to one and-half inches long in the skin with a knife. The doctor then uses sutures or stitches to close these cuts at the end of the procedure. In the no-scalpel vasectomy, the doctor makes only one tiny puncture into the skin with a special instrument instead of making two incisions. This same instrument is used to gently stretch the skin opening to reach the tubes easily.

The tubes are then blocked, using the same methods as conventional vasectomy. Because a scalpel is not used, there is minimal bleeding. Usually, no stitches or sutures are needed to close the tiny opening. This opening will heal quickly with little or no scarring. No-scalpel vasectomy was introduced in the United States in 1988. Many doctors who have mastered the no-scalpel technique now use the no-scalpel technique almost exclusively.

"How Long Does the No-Scalpel Vasectomy Take?"

It depends on the surgeon, but on average the operation takes between fifteen to thirty minutes.

Primary Benefits of the No-scalpel Vasectomy

1. No incision with a scalpel—only a small opening
2. Usually, no stitches or sutures will be necessary as the opening is less than ¼ inch
3. Usually, a faster procedure
4. Usually, a faster recovery
5. Usually, less chance of bleeding and other complications
6. Usually, less discomfort
7. Just as effective as the regular vasectomy

Questions and Concerns Before a Vasectomy

"Will My Vasectomy Hurt?"

When the local anesthetic is injected into the skin of the scrotum, you will feel some mild discomfort, but as soon as the anesthetic takes effect, you should feel no pain for the rest of the procedure. We recommend that men plan to go home after the procedure and place an ice pack on the scrotum for a few hours to reduce any swelling. Many men will elect to use a bag of frozen peas that will conform to the scrotum, but any ice pack or plastic Zip-Loc® bag with a few ice cubes will work just fine. We suggest that the ice pack be placed on the underwear and not directly on the skin.

We also recommend wearing jockey underwear and not boxer underwear for a few days after the procedure to keep your "package" lightly compressed. Afterward, you will experience mild discomfort for a couple of days. You may want to take an over-the-counter painkiller such as Tylenol. Still, the discomfort is usually less with the no-scalpel technique because of less trauma or injury to the scrotum and tissues surrounding the vas. Also, there are no stitches or sutures in most cases. Your doctor will provide you with complete instructions about what to do after surgery.

"How Soon Can I Go Back to Work?"

You should be able to do routine physical work within forty-eight hours after your vasectomy. You will be able to do heavy physical labor and exercise within a week.

"Will the Vasectomy Change Me Sexually?"

The only thing that will change is that you will not be able to make your partner pregnant. Your body will produce the same hormones that give you your sex drive and maleness. You will make the same amount of semen. A vasectomy will not change your beard, muscles, sex drive, erections, climaxes, or even raise your voice! Some men say that without the worry of accidental pregnancy and the bother of other birth control methods, sex is more relaxed and enjoyable than before.

"Will I Be Sterile Right Away?"

No! No! No! We say this three times to emphasize that there are some active sperm located above the area where the vas is divided after a vasectomy. It may take a dozen to two dozen ejaculations to clear the sperm from where the vasectomy is performed. You and your partner should use other forms of birth control until specimens are examined on two separate occasions to ensure no sperm are present.

"Is Either Type of Vasectomy Safe?"

Vasectomy is safe and straightforward. Vasectomy is an operation, and every surgery has some risks, such as bleeding, infection, and pain, but serious problems are unusual. There is always a tiny chance of the tubes rejoining themselves, which is why sperm checks are necessary. There have been some controversies about the long-term effects of vasectomy. Still, to our knowledge, there are no long-term risks to vasectomy.

"When Can I Start Having Sex Again?"

As a rule, we suggest waiting a week before having intercourse. Remember, however, that the vasectomy only divides the vas and has no effect on the sperm that are already beyond that point. It is essential not to have unprotected intercourse until the absence of sperm from the ejaculate has been confirmed with two negative sperm checks. In other words, keep using your usual form of birth control until you have been given the "all clear" by your urologist.

"Can I Reverse My Vasectomy if I Change My Mind?"

The decision to opt for a vasectomy remains a highly personal one. The potential risks and benefits must be considered, including the possibility that you may change your mind. Vasectomy reversal is possible, but success is not guaranteed and depends mainly on how long ago it was done. So, it is much better to consider it a permanent procedure. Also, a vasectomy reversal can be costly. If you want a vasectomy but may later want kids, consider freezing sperm before the surgery. Your doctor can usually help you locate a lab that freezes and stores sperm.

Both vasectomy reversal (VR) and in vitro fertilization (IVF) represent alternative options for the couple seeking fertility after vasectomy. Specific circumstances may favor one modality over another, depending on the interval between the vasectomy and when the couple wishes to achieve pregnancy. Other factors impacting a successful pregnancy include female fertility factors, female partner age, male partner age, and cost. In the absence of insurance coverage, vasectomy reversal is often more cost-effective than IVF. Alternatively, when a female factor contributes to infertility, IVF is often the better choice.[1] Pregnancy rates for both options are similar. Making a final choice through shared decision-making while considering these options is ideal.

"What is the Cost of a Vasectomy?"

Costs range from several hundred to several thousand dollars. You should ask the doctor if the cost includes:

- The initial consultation.
- The cost of the procedure.
- The post-procedure semen examinations.

The Future of Male Sterilization

Several hormonal male contraceptive agents are currently under investigation. Novel single agent products evaluated include dimethandrolone undecanoate, 11β-methyl-nortestosterone dodecylcarbonate, and 7α-methyl-19-nortestosterone. A contraceptive efficacy trial of Nestorone®/testosterone gel is underway. There is a potential for non-hormonal methods, which are at the preclinical stages of development. Many non-hormonal male contraceptive targets that affect sperm production, sperm function, or sperm transport have been identified. They may be ready for clinical trials in the future.[2]

What Happened to Paco?

Paco had counseling with a urologist. Both he and his partner discussed having a vasectomy. He and his wife talked it over and decided to proceed with the vasectomy. Paco had a no-needle, no-scalpel vasectomy in the doctor's office. The procedure took fifteen minutes, and he left the office, drove home, and used an ice pack on his scrotum until he went to sleep. He refrained from exertion and sexual intimacy for one week. When Paco brought two semen samples to the doctor's office to be examined under a microscope, no sperm was found in either sample. He was declared sterile, and he was free to engage in sexual intimacy without any need for contraception. His libido and erections were unchanged, and he and his partner were happy campers. They told many of their friends about their positive vasectomy experience.

Bottom Line

A vasectomy is one of the best ways to achieve contraception. It is nearly 100 percent successful and can be done in the doctor's office under a local anesthetic without using a needle or a scalpel. The procedure is accomplished in just a few minutes, and most men have no pain or discomfort afterward. This truly is the "prime cut!"

References

1. Dubin, Justin M., Joshua White, Jesse Ory, and Ranjith Ramasamy. "Vasectomy Reversal vs. Sperm Retrieval with In Vitro Fertilization: A Contemporary, Comparative Analysis. *Fertility and Sterility* 115, no. 6 (June 2021): 1377–83. https://doi.org/10.1016/j.fertnstert.2021.03.050.
2. Dominiak, Zuzanna, Hubert Huras, Paweł Kręcisz, Waldemar Krzeszowski, Paweł Szymański, and Kamila Czarnecka. "Promising Results in Development of Male Contraception." *Bioorganic & Medicinal Chemistry Letters* 41 (June 2021): 128005. https://doi.org/10.1016/j.bmcl.2021.128005.

CHAPTER 7

Battling Low Testosterone

Patrick deals with the same stresses that most men in their mid-fifties experience—long work hours, financial strain, and a little weight gain. Patrick attributed his ongoing fatigue to these factors. However, his wife, Nancy, noticed that Patrick would spend most of his evenings and weekends on the couch over the last year. He no longer had the energy for golf and other exertional activities. His wife also noticed that his mood seemed unpredictable. When Patrick became increasingly inattentive in the bedroom, Nancy knew something was wrong. She convinced Patrick to make that long-overdue appointment with his primary care physician.

What Is Low T?

Low testosterone, commonly referred to as "low T," is a condition that has recently exploded in the public eye. The onslaught of self-diagnoses has been fueled by the plethora of new testosterone replacement products and the vague definition of this ailment. In fact, most low testosterone symptoms are also shared with other more common medical conditions. For instance, a low energy level has dozens of explanations, with low testosterone being one of the less likely causes. Nevertheless, testosterone blood levels decrease by 1 percent annually starting at age thirty. More than one in five men will have low testosterone levels by age sixty.[1]

Most doctors agree that replacement should only be considered in patients with a low blood testosterone level *and* at least one symptom associated with low testosterone. For instance, if a man's serum testosterone level is normal, testosterone should not be administered to treat fatigue or sexual dysfunction. Your physician should search for other causes of your symptoms before considering testosterone replacement, especially if the testosterone level in the blood is normal.

Physicians also have some disagreement on what blood levels are considered normal. The normal range of testosterone is 300-900 ng/dL. Typically, any blood level below 300 ng/DL is low. However, this level can vary based on other factors, such as the time of day that the blood is drawn. It is generally recommended that the blood test be obtained in the morning when testosterone levels peak. If a low testosterone level is obtained late in the day, the test should be repeated in the morning. Repeat testing is unnecessary if the afternoon blood draw is in the normal range.

For borderline testosterone levels, additional testing should be performed. The standard test we discussed is called a "total testosterone" level. This measures free-testosterone and testosterone bound to proteins in the bloodstream. The "free" testosterone is the active component and the hormone responsible for many bodily functions, including libido, bone strength, muscle mass, energy level, and even a man's mood. The typical symptoms associated with low testosterone include low sex drive (libido), decreased erectile function, loss of muscle mass, reduced bone mass (osteoporosis), mood changes, depression, increased body fat, falling asleep after meals, and fatigue. Although there are ways to either calculate or directly measure free testosterone levels, this test to measure free testosterone is expensive and has fallen mostly out of favor. Measuring the amount of bioavailable testosterone, the testosterone bound to proteins in your blood, is the best test for borderline patients.

So, back to the question: What is low T? Some say it is merely a low testosterone level. Others say that symptoms usually associated with low testosterone levels meet the criteria, even if the blood levels are not decreased. The name "low T" renames the well-established medical condition, hypogonadism. It is not just a chemical (blood test) diagnosis, nor is the term just a clinical diagnosis. Hypogonadism (androgen deficiency) is when the testicles do not produce the proper amount of testosterone. As a result, clinical symptoms consistent with testosterone deficiency develop.

Primary Causes of Low T

The most common cause of androgen deficiency is the pituitary gland's decreased production of luteinizing hormone, LH (Figure 1). With the reduction in LH from the pituitary gland, the testicles are not stimulated to produce testosterone. In fact, when testosterone levels are low, the pituitary gland should have high levels of LH to stimulate the testicles to produce more testosterone. The decrease in LH production in middle-aged and older men occurs for unknown reasons. This condition is called secondary hypogonadism. On rare occasions, this can be caused by a benign pituitary gland tumor that can only be diagnosed with a special imaging scan, magnetic resonance imaging, or MRI.

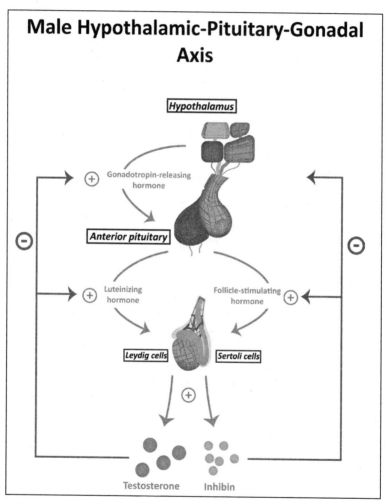

Figure 1: The Pit Bosses, the Balls

Primary hypogonadism occurs when the testicles are damaged and unable to produce adequate testosterone. Primary hypogonadism includes:

- Undescended testicles (see Chapter 8)
- Prior testicular infections
- Chemotherapy
- Previous inguinal or scrotal surgery

Other causes of low testosterone disturb the interaction between the pituitary gland and the testicles. These include diabetes, chronic renal disease, obstructive sleep apnea, use of opiates, use of certain antidepressants (Celexa®, Lexapro®, Prozac®, Paxil®, Zoloft®), drugs used to treat prostate cancer, chronic liver failure, alcoholism, drug abuse, and depression.

The Risks of Testosterone Replacement

What would be the problem with trying testosterone replacement and seeing if it works? Some men with "normal" testosterone levels may feel their level might not be normal. The first problem with that approach is that a man with the right testosterone level can feel better with an additional amount. In fact, for this reason, testosterone can be very addictive. One common example is the bodybuilder that uses excess doses of testosterone to build muscle without any medical oversight. This addictive nature caused the FDA to regulate testosterone as a controlled substance, like narcotic medications. The FDA does not permit prescribing testosterone over the phone. A written prescription must be obtained from the doctor in most states.

Testosterone replacement can also affect your body's own natural testosterone. The pituitary gland is a small gland located below the brain that regulates how certain hormones are produced. Suppose the pituitary gland detects testosterone in the blood from *any* source. In that case, it will suppress the testicles to stop producing testosterone. With this concept in mind, testosterone replacement is just that—"replacement." Otherwise, it would have been called testosterone "enhancement." The longer a man receives testosterone replacement therapy, the longer it will take for his natural production to return once the replacement is stopped.

The pituitary gland also affects sperm production in the testicle. The pituitary gland will stop sending these hormonal signals to the testicle when receiving testosterone. As a result, the production of sperm and the quality of the sperm will plummet. This effect on fertility is usually reversible. However, for men interested in maintaining their fertility, the pituitary hormone HCG is given as a subcutaneous injection, or Clomid, an oral medication, is used instead of testosterone. Clomid and HCG will cause the man's testicles to produce more testosterone, not adversely affecting sperm production.

Too much testosterone can cause the overproduction of red blood cells. This is an example of "too much of a good thing." Our blood becomes too thick if the red blood cell count or hematocrit rises above 50 percent. This excess blood may clog small arteries—such as those going to the heart or brain—thereby increasing the risk of heart attack or stroke. This condition of excess red blood cells is different from blood clot formation. Most men with too many red blood cells can be managed by donating blood at a blood bank and repeating the red blood cell count a few weeks later.

Testosterone also affects the prostate. It will potentially increase the prostate size, disturb urinary habits, and raise the PSA levels (blood test for prostate cancer screening; see Chapter 3.) Although there is no evidence that testosterone causes prostate cancer, careful monitoring of the PSA level and a digital examination is recommended during testosterone replacement. Most doctors recommend that men receiving testosterone replacement have a PSA and digital rectal exam every six to twelve months.

Several reports have shown that testosterone replacement increases the risk of cardiovascular events.[2] However, these studies involve men with other cardiovascular risk factors. Careful monitoring should be conducted on any man considered for testosterone replacement. In men at a higher risk for cardiovascular disease, replacement should be avoided despite any androgen deficiency symptoms or low blood testosterone levels. Monitoring should include regular blood counts, testosterone levels, PSA blood tests, and liver function tests. Men should only receive the minimum testosterone necessary to achieve essential goals.

The Risks of Low Testosterone

Testosterone is an essential hormone. It is partly responsible for maintaining lean body mass, bone health, and red blood cell production. Testosterone deficiency increases the risk of obesity, diabetes, metabolic syndrome, osteoporosis, and clinical depression. Interestingly, weight loss, diabetes control, moderate exercise (especially weightlifting), better sleeping habits, and even sex can significantly improve testosterone levels. This is an excellent example of "the chicken and the egg." Is the deficiency causing the symptoms, or are the symptoms causing the deficiency? It's probably both.

The Cardiovascular Effects of Testosterone Replacement

Testosterone replacement in certain patients can increase the risk of cardiovascular events such as stroke and heart attack. It is also true that men with androgen *deficiency* are more likely to experience those difficult cardiovascular events.[3]

What are the explanations behind the seemingly contradictory ends of the spectrum? With testosterone deficiency, for example, obesity is more common—and certainly not heart-healthy. However, patients with existing heart disease are more sensitive to changes in their blood volume. Testosterone replacement can cause fluid retention and place additional stress on the heart.

Every man has an ideal range for testosterone levels. Too much testosterone is not good, nor is too little desirable. This concept applies to many of our bodily demands. Your doctor's goal is to help you get to the testosterone level that addresses your low T and its unfortunate symptoms.

A Note for Prostate Cancer Patients and Those at Risk for Prostate Cancer

Testosterone replacement has never been shown to cause prostate cancer. However, special care and monitoring should be given to men at increased risk for developing prostate cancer: African Americans; men with a family history of prostate cancer; men with elevated PSA blood tests; men with an abnormal digital rectal exam. However, the same rules apply to men wishing to receive testosterone replacement therapy: low testosterone blood level and significant symptoms consistent with androgen deficiency.

A urologist's input is essential for men treated for prostate cancer. Testosterone replacement is reasonable for low- to medium-risk prostate cancer patients who have seemingly been successfully treated. The goal is to put men with androgen deficiency on the same playing field with the same energy level and sex drive as the average guy with a normal T level, albeit with careful dosing and monitoring. After all, for the average man with normal testosterone production, we would not put him on medications to lower his testosterone.

We should not deny treatment for low T in men seemingly cured of prostate cancer. Men treated for prostate cancer, either with radiation therapy or surgical removal of the prostate gland, must have a PSA level that has remained low for several months before initiating testosterone replacement. These men must commit to regular PSA testing. If the PSA increases following testosterone replacement therapy, the treatment for low T must be stopped immediately. Even in the worst-case scenario, the PSA will eventually begin to rise and reveal the residual disease in a patient not cured of prostate cancer. The testosterone replacement would then cease, and the urologist would initiate additional treatment for the prostate cancer.

Initiating Treatment

Your physician will conduct a detailed history and physical exam to confirm the diagnosis described above and possibly determine the underlying cause. Repeated blood tests are often necessary. In some cases, these tests would include a testosterone level, a PSA level, and pituitary hormones. Your medical history will also help assess the risks associated with hormone deficiency and replacement.

Methods of Administering Testosterone

Until recently, oral medications were usually not an option. In the past, testosterone pills had a high likelihood of damaging the liver compared to other routes of administration. Certainly, the oral administration of testosterone is the more preferable choice. An oral preparation of testosterone undecanoate (Tlando), one tablet twice a day, has been shown to restore testosterone levels to normal. Men who use oral preparation report improvement in their libido and sexual performance. Oral testosterone may improve patient compliance because of its ease and convenience.[4]

Topical dosing through the skin comes in three forms: patch, gel, and underarm liquid. These topical treatments require daily morning application and mimic natural testosterone production better than other replacement methods. However, some men find these topical testosterones inconvenient, messy, and easy to forget to use every day. Care must be taken not to transfer the drug to another person's skin (especially children). We strongly recommend that all men carefully wash their hands after applying the testosterone gel. Cost can also be very high, although this can often be remedied in part by using a compounding pharmacy, where medications are actually formulated onsite and specifically tailored to the customer's needs.

Intramuscular testosterone injections require an injection with a needle every one to two weeks. This treatment option is the most common method of testosterone delivery, probably because it has been around the longest. However, it isn't easy to maintain consistent levels. For instance, during the first few days following the injection, the levels can be very high and even cause euphoria. Conversely, in the days leading up to the next injection, the level can fall precipitously and cause a significant downswing in mood and energy. We often refer to this as the "yo-yo effect."

A novel method uses a tiny wafer placed at the gum line on the inner cheek. This is called a buccal patch. However, most men dislike keeping something in their mouths for long periods.

A longer-lasting option involves the implantation of testosterone pellets below the skin, usually in the buttocks area. These small rice-sized pellets typically last six months. However, as with the injections, the initial levels can be too high, and the later levels (the last month) can be too low. It is also difficult to reverse, as removing the pellets once inserted under the skin is almost impossible. Pellets should not be used in men with prostate cancer since it would be difficult to remove the small pellets if the PSA were to rise.

Testosterone patches are called transdermal patches and are applied to the skin. The testosterone in the patch, like the topical gels, will pass through the skin and enter the bloodstream, where it can supply the organs and body parts that need testosterone. Patches are administered and replaced once a day. The patches are worn on the back, upper arm, thigh, or abdomen. One brand of testosterone patch is applied to the scrotum. The most common problem with these patches is skin irritation with visible red blotches that are often itchy or uncomfortable. It may take several weeks before men will start to feel the beneficial effects of the testosterone patch.

Side Effects of Testosterone Treatment

Testosterone replacement therapy is relatively safe, and side effects are not common. Most of the side effects are transient and can be resolved by reducing the dosage or, if necessary, discontinuing the use of testosterone.

Testosterone therapy may raise a man's risk for blood clots and stroke; it may result in secondary erythrocytosis—too many solid red blood cells in relation to the liquid plasma, It's an unfortunate result of the stimulation of red blood cell production. We recommend that all men on testosterone replacement therapy undergo a complete blood count (or CBC) test every six months. If the red blood cell count is significantly increased, then men are advised to donate blood at a blood bank and have the blood count repeated in a few weeks to see that the count is in the normal range. The testosterone dosing can be lowered to achieve an acceptable red blood cell count. However, regular blood donation will become a part of the treatment in some men.[5]

Testosterone administration suppresses the follicle-stimulating hormone (FSH) that is produced in the pituitary gland. As a result, there is a decrease in sperm production. Therefore, men interested in future fertility should be cautious about using testosterone. On occasion, the reduction in sperm count may be permanent and preclude a pregnancy.[6] An alternative for men with low testosterone who wish to preserve their fertility would be as candidates for receiving human gonadotropic treatment to increase testosterone level and sperm production.[7]

Testosterone replacement therapy may exacerbate obstructive sleep apnea (OSA). Men with a history of OSA should be aware of this side effect and report any symptoms of increased snoring, brief periods of breathing cessation, and chronic fatigue.[8]

On occasion, men who use testosterone replacement for extended periods may report a reduction in the size of their testicles. This side effect can be reversed if testosterone is discontinued.[9]

Men on testosterone replacement therapy may develop breast enlargement, gynecomastia. (See chapter 13.) Gynecomastia may be accompanied by nipple tenderness and is resolved by reducing the testosterone dose.[10]

Testosterone can also result in facial acne. This is more common in younger men and usually responds to topical skin therapy and reduced testosterone dosage.[11]

Earlier in this chapter, we advised that men who use a topical testosterone gel must wash their hands thoroughly after applying topical testosterone and also watch that no one touches the skin area where they have applied the medication. If a woman or child came in contact with testosterone gels, it could transfer side effects to them, including hair growth in women and premature puberty in young children.[12]

Contraindications for Testosterone Replacement Therapy

Testosterone replacement is contraindicated in men with breast cancer or untreated prostate cancer, high red blood cell count, congestive heart failure (CHF), liver disease, and significant water retention. Caution should also be taken in men with sleep apnea, benign prostatic enlargement, and cardiovascular disease.

The Benefits of Testosterone Replacement

Despite the potential perils, testosterone is an essential hormone. A return to a normal level can lead to the following benefits: enhanced libido, increased muscle mass, decreased fat mass, better control of diabetes, correction of anemia, improvement in bone strength, mood improvement, sense of well-being, and improvement in cholesterol and lipid levels.

Andropause

Ironically, the word *menopause* contains the word *men*. Well, andropause is the male equivalent of menopause. Like many topics in this chapter, the concept of andropause is controversial. Testosterone levels naturally decline with each decade of life. Some argue that this is a normal process that should not be treated. However, no well-established blood testosterone levels have been adjusted for age.

The Androgen Deficiency in the Aging Male (ADAM) questionnaire[13] concerns low testosterone symptoms. It was developed to help men describe the kind and severity of their low testosterone symptoms. Here are the ten questions used to evaluate for low testosterone:

1. Do you have a decrease in libido (sex drive)?
2. Do you have a lack of energy?
3. Do you have a decrease in strength and/or endurance?
4. Have you lost height?
5. Have you noticed a decreased "enjoyment of life?"
6. Are you sad and/or grumpy?
7. Are your erections less strong?
8. Have you experienced a recent deterioration in your ability to play sports?
9. Are you falling asleep after dinner?
10. Has there been a recent deterioration in your work performance?

If you answer "yes" to number 1 or 7 or if you answer "yes" to more than three questions, you may have low testosterone.

Like most approaches in medicine, treatments should be individualized. Testosterone replacement should not be considered the fountain of youth. Testosterone replacement will turn back the clock less than every man would like. The administration should be targeted toward reducing the harmful health effects of declining testosterone production.

What Happened to Patrick?

After a complete evaluation, his physician determined that he met all the criteria for clinical androgen deficiency. He started on a daily testosterone gel preparation, and his testosterone blood level went from 150 dL/mL to 450 dL/mL. Although not perfect, his mood, energy level, and libido dramatically improved. Nancy does not describe him as "a new man." She describes him as "the man she always knew." And Patrick appreciates his rejuvenated self!

Bottom Line

Testosterone deficiency affects millions of middle-aged and older men. Many men with low T levels will have symptoms such as low energy levels, decrease in muscle mass, decrease in bone mineral density, and a more negative mood. The diagnosis of low T is easily made with a blood test best obtained in the morning. Treatment options for low T include injections, topical testosterone, and testosterone pellets. Most men with low T and symptoms of low T are candidates for testosterone replacement therapy. Men receiving testosterone replacement should have regular checkups for their PSA levels and red blood cell counts. If you aren't feeling your manhood as much as you would like, then speak to your doctor and have a testosterone blood test.

References

1. Brawer, Michael K. "Testosterone Replacement in Men with Andropause: An Overview." *Reviews in Urology* 6, suppl. 6 (2004): S9–S15. https://www.ncbi.nlm.nih.gov/pmc/articles/PMC1472881/.
2. Corona, G. Giovanni, Giulia Rastrelli, Elisa Maseroli, Alessandra Sforza, and Mario Maggi. "Testosterone Replacement Therapy and Cardiovascular Risk: A Review." *World Journal of Men's Health* 33, no. 3 (December 2015): 130–42. https://doi.org/10.5534/wjmh.2015.33.3.130.
3. Morris, Paul D., and Kevin S. Channer. "Testosterone and Cardiovascular Disease in Men." *Asian Journal of Andrology* 14, no. 3 (May 2012): 428–35. https://doi.org/10.1038/aja.2012.21.
4. DelConte, Anthony, Kongnara Papangkorn, Kilyoung Kim, Benjamin J. Bruno, Nachiappan Chidambaram, Mohit Khera, Irwin Goldstein, Tobias S. Kohler, Martin Miner, Adrian S. Dobs, and Mahesh V. Patel. "A New Oral Testosterone (TLANDO) Treatment Regimen without Dose Titration Requirement for Male Hypogonadism. *Andrology* 10, no. 4 (May 2022): 669–76. https://doi.org/10.1111/andr.13153.
5. White, Joshua, Francis Petrella, and Jesse Ory. "Testosterone Therapy and Secondary Erythrocytosis." *International Journal of Impotence Research* 34, no. 7 (November 2022): 693–97. https://doi.org/10.1038/s41443-021-00509-5.
6. Sharma, Aditi, Suks Minhas, Waljit S. Dhillo, and Channa N. Jayasena. "Male Infertility Due to Testicular Disorders." *Journal of Clinical Endocrinology & Metabolism* 106, no. 2 (February 2023): e442–e59. https://doi.org/10.1210/clinem/dgaa781.
7. Fink Julius, Brad J. Schoenfeld, Anthony C. Hackney, Takahiro Maekawa, and Shigeo Horie. "Human Chorionic Gonadotropin Treatment: A Viable Option for Management of Secondary Hypogonadism and Male Infertility." *Expert Review of Endocrinology & Metabolism* 16, no. 1 (January 2021): 1–8. https://doi.org/10.1080/17446651.2021.1863783.

8. Payne, Kelly, Larry I. Lipshultz, James M. Hotaling, and Alexander W. Pastuszak. "Obstructive Sleep Apnea and Testosterone Therapy." *Sexual Medicine Reviews* 9, no. 2 (April 2021): 296–303. https://doi.org/10.1016/j.sxmr.2020.04.004.

9. Palacios, A., R. D. McClure, A. Campfield, and R. S. Swerdloff. "Effect of Testosterone Enanthate on Testis Size." Journal of Urology 126, no. 1 (July 1981): 46–48. https://doi.org/10.1016 /S0022-5347(17)54372-4.

10. Corpas, Emiliano, Marc R. Blackman, S. Mitchell Harman, Ricardo Correa, and Antonio Ruiz-Torres. "Male Hypogonadism in Advanced Age (Therapeutic Considerations), Gynecomastia in Advanced Age, Benign Prostatic Hypertrophy and Prostate Cancer (Endocrinological Aspects of Development and Treatment)." In *Endocrinology of Aging: Clinical Aspects in Diagrams and Images*, 407–32. Amsterdam: Elsevier, 2021.

11. Kumtornrut, Chanat, and Nopadon Noppakun. "Androgens and Acne." In *Acne: Current Concepts and Management*, edited by Dae Hun Suh, 179–87. Updates in Clinical Dermatology. Cham, Switzerland: Springer, 2021.

12. Stahlman Jodi, Margaret Britto, Sherahe Fitzpatrick, Cecilia McWhirte, Samuel A. Testino, John J. Brennan, and Troy L. Zumbrunnen. "Effect of Application Site, Clothing Barrier, and Application Site Washing on Testosterone Transfer with a 1.62% Testosterone Gel." Current Medical Research and Opinion 28, no. 2 (February 2012): 281–90. https://doi.org/10.1185/030079 95.2011.652731.

13. Tancredi, Annalisa, Jean-Yves Reginster, Florence Schleich, Georges Pire, Philippe Maassen, Francoise Luyckx, and Jean-Jacques Legros. "Interest of the Androgen Deficiency in Aging Males (ADAM) Questionnaire for the Identification of Hypogonadism in Elderly Community-Dwelling Male Volunteers." *European Journal of Endocrinology* 151, no. 3 (September 2004): 355–60. https://doi.org/10.1530/eje.0.1510355.

CHAPTER 8

Scrotal Issues and Testicular Pain

Pomeroy's wife was flipping through her husband's recent issue of Men's Health *when she noticed an article about men checking their bags. And no, she was not reading the travel section. Of course, she left it open to that page and left it on the kitchen counter for her Pomeroy to read. The next morning, Pomeroy discovered by self-examination a small and tender lump "down there." Pomeroy could not make an appointment with the urologist quickly enough.*

Scrotal Anatomy: How Is That Bag Packed?

Most men do not pay attention to what's going on down there until there is a problem. However, some problems down there do not cause any symptoms. It is also important to be familiar with your particular anatomy in order to recognize any abnormal changes.[1]

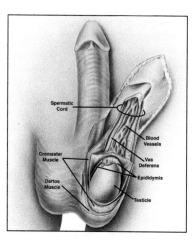

Figure 1: The Scrotum

The skin of the scrotum is very specialized. As you probably know, the testicles are located outside of the body in order to keep the family jewels at the proper temperature. Normal body temperature is a bit too balmy for those freshly made sperm. The scrotal skin contains the dartos muscle that raises and lowers the testicles closer and further from the body in order to regulate their finicky temperature needs. During this process, the skin will wrinkle as the scrotum is drawn closer to the body.

The scrotum contains a lot more than two balls. Behind each testicle is a structure called the epididymis. It is a smooth but soft coil of small tubes that store sperm previously manufactured in the testicles. The sperm exit the epididymis through a tube called the vas deferens (or "vas" as in "vasectomy"). The vas travels into the abdomen through the groin (also called "inguinal canal"). Alongside the vas are the testicular artery and veins. Collectively, the vas tube and these blood vessels are referred to as the spermatic cord. This spermatic cord also contains the cremaster muscle which aids the dartos muscle in regulating the scrotal temperature and keeping the testicles a degree or two cooler than the core temperature of the body of 98.6°F. The cremaster also contracts in response to any physical threat to the scrotum such as a swift kick, causing the testicles to rise and stay out of harm's way.

The testicles basically perform two functions: sperm production and testosterone production. Anything that can damage the testicles can also affect these two functions. Both of these functions are regulated by hormones produced by the pituitary (a small gland located just below the brain).

A Problem in the Sack

No, we're not talking about poor performance in the bedroom! However, problems in the scrotum can lead to issues with sexual prowess and fertility. We will spend the remainder of this chapter discussing such problems as pain, abnormal swelling, and even cancer. And, of course, we'll delve into the solutions for these problems.

Scrotal Pain

The two most common causes of scrotal pain are infection and trauma. On rare occasions, a prior vasectomy or hernia repair can cause chronic pain. However, sometimes the cause is not so obvious. In some cases, ball pain can become a long-term and difficult to treat problem.

The first step in evaluating testicular pain is differentiating whether it is tender or just painful. If it is tender, the problem usually originates in the testicle or epididymis (such as an infection, covered later in this chapter). However, if it is painful but not tender to the touch, the pain could be coming from somewhere else. This condition is called *referred pain*. A common example of referred pain occurs during a heart attack, during which the pain "radiates" to the left arm. Of course, the heart is not located in the arm. Similarly, the painful passage of a kidney stone can cause pain in the testicle. This phenomenon occurs because the sensory nerves of the kidney enter the spinal cord in a similar location to the testicular nerves. Other examples of referred testicular pain include inguinal hernia, prior hernia repair, and prostate problems.

A careful history and physical exam can often determine the underlying cause. In other cases, an ultrasound or CT scan may be helpful. When the cause cannot be determined or treated, therapy is directed towards the pain itself. Anti-inflammatory, anti-spasmodic, and anti-seizure (for example, Neurontin®) medications have all been tried with varying success.

A pain specialist physician may play a role with such treatments as a nerve block. This procedure is initially performed with a local anesthetic injection in the nerves leading to the testicle. If this temporary treatment is successful, a longer-lasting destruction of the nerve could be performed using radiofrequency (heat) energy.

Another option for controlling chronic pain is implantation of an electrical nerve stimulator (Interstim™ device). As a last resort, the nerves leading to the testicle can be surgically cut. Although this procedure can be performed with a laparoscopic technique, the nerves must be transected (cut across) in the back portion of the abdomen.

With any of these procedures, it is not unusual for the pain to return within one to two years. In any case, a man with such a severe case of testicular pain must be his own advocate and seek advice from multiple specialists until he finds an answer. In its extreme form, it can be very difficult for a man and his loved one to cope with this condition.

The Groin Hernia

The body is made of many compartments, and every structure is assigned to its own compartment. When a part of our anatomy either can or does go from its own compartment into another compartment, it is called a hernia. For instance, if the stomach protrudes through the diaphragm and into the chest, it is called a hiatal hernia. The most common type of hernia is an inguinal, or groin, hernia. In this case, an abdominal structure such as the intestine can protrude from the abdomen into the scrotum or into the area just above the scrotum (the inguinal canal).

The most common symptom from an inguinal hernia is testicular or groin pain. Often, there are no symptoms, but a man or his physician notices a bulge in the area. In most cases, a groin hernia requires surgical correction. If left untreated, a portion of the intestine can become trapped outside of the abdomen leading to a surgical emergency.

A hernia repair is typically a safe, outpatient procedure. A man can usually return to work in a couple of days following the procedure, and he can typically resume a light exercise routine within one to two weeks. For most hernia surgeries, a piece of permanent synthetic mesh is used for the reconstruction. Problems with the use of mesh are rare, but there are reported cases of entrapment of the testicular nerve causing long-term pain. This complication would possibly require removal of the mesh in order to release the nerve.

Extra Parts: Cysts, Hydroceles, and Varicoceles

Some people are blessed with bigger parts, and some are blessed with extra parts. Cysts are fairly common in the scrotum. A cyst is merely a thin-walled sphere containing fluid. They come in a dramatic variation of sizes. When located in the epididymis, adjacent to the testicle, they are often called spermatoceles since they contain sperm. No treatment is required unless they cause pain.

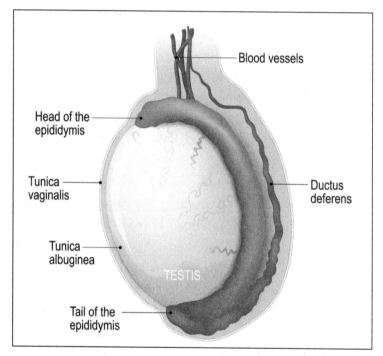

Figure 2: Anatomy of the Testicle

Surrounding the testicle is a thin sac that contains a small amount of fluid which lubricates the testicle while it bounces around. If this fluid builds up, it becomes a hydrocele. Like the spermatocele, no treatment is required unless it causes pain. However, if a hydrocele contains so much fluid that the testicle cannot be felt, it should be evaluated by a urologist.

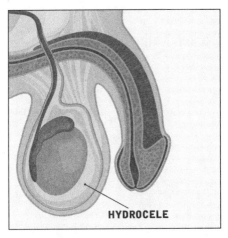

Figure 3: Hydrocele *(Shutterstock)*

If you feel a "bag of worms" down there, you may have a varicocele. A varicocele is the testicular version of varicose veins that are common in the legs. With this condition, the little valves within the veins do not work properly, and the blood backs up. As a result, the veins can become very dilated. When you lie down, the blood drains better and the veins become smaller. However, you should still see your doctor to make the final diagnosis. Because of our internal anatomy, varicoceles are much more common on the left. In fact, if it is only present or more predominant on the right, an abdominal tumor could be the underlying cause.

Most varicoceles do not cause symptoms, but some men will have pain. Also, the additional blood in the scrotum raises the temperature down there and could cause fertility problems. The fix is a minor surgical procedure in which the urologist ties off the veins through a small incision above the scrotum.

Your physician can usually diagnose those extra parts down there with a simple physical exam. A scrotal ultrasound can confirm the diagnosis.

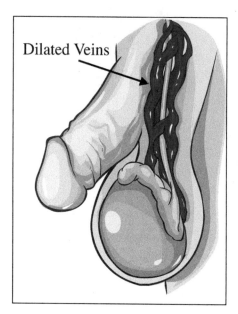

Figure 4: Varicocele

Testicular Trauma

Most men have experienced that very distinct pain associated with a foot or projectile to the crotch. Since the nerves from the testicles enter the spinal cord in a nearby location to the intestines, severe nausea can quickly follow trauma to the jewels. In some cases, the pain can be so intense that it causes a man to black out. Most injuries resolve on their own with nothing more than a little tender loving care; if the testicle were to rupture, however, it becomes a surgical emergency. Any significant swelling or bruising should raise this concern. The diagnosis is easily made with an ultrasound of the scrotum. Failure to repair a ruptured testicle often leads to loss of that precious jewel.

Serious injuries to the testicle can also damage other parts. Any scrotal skin lacerations should be promptly treated. The urethra (the tube through which urine exits the bladder) passes just behind the scrotum prior to traveling though the penis. The urethra can sustain significant damage if the testicle is injured by a "straddle" injury (such as slipping off a bike seat and landing on the crossbar with a leg on either side.) Blood at the tip of the penis or in the urine is one of the indicators of a urethral injury. Sometimes, a man will not notice the signs of urethral trauma in the face of severe testicular pain. As always, when in doubt, check it out.

The Skinny on Scrotal Skin

Almost anything that can affect the skin elsewhere can affect the scrotal skin. Because of the predominance of hair follicles and sweat glands in a bacteria-rich environment, sebaceous cysts are very common down there. These are round nodules contained in the scrotal wall, separate from any of the internal structures described earlier. Usually, these should be removed to prevent any future problems.

Jock itch is also a common scrotal skin problem. The red rash and itching can sometimes be painful. This fungal infection tends to affect the moister area between the scrotum and the thigh. Careful hygiene and an antifungal cream or spray powder usually eliminates the problem.

Warning: If you notice any infected or tender pimple or boil on (or near) the scrotum, seek medical attention immediately. These infections can spread rapidly, destroy large amounts of skin and muscle, and become life-threatening—especially in diabetics or immunocompromised men. For those who choose to remove hair down there, clippers are far superior to razors to prevent ingrown hairs and subsequent skin infections.

Scrotal Edema

The scrotum can often serve as a pop-off valve for the body. Any fluid buildup in the area can easily cause the scrotum to expand to several times its normal size. Any fluid or blood from abdominal surgery will easily travel to the scrotum.

The good news is that the scrotum will return to normal with time. Elevating your swollen sack with a folded hand towel may help reduce the size. The swelling will not restrict urine flow, but it can make your aim even worse.

Bacterial Infections

Bacteria from the genitourinary or gastrointestinal tracts can make their way into the scrotal sack and cause an infection in the epididymis (epididymitis), testicle (orchitis), or both (epididymoorchitis). The testicle can become very painful, hard, and swollen. In more severe cases, bacteria can spread to the bloodstream and cause high fevers. The diagnosis is usually clinical, but it can be confirmed with a scrotal ultrasound. It is important to distinguish this diagnosis from testicle torsion (discussed below).

Some cases require two to three weeks of antibiotics, yet the symptoms may take over a month to resolve. Nonsteroidal anti-inflammatory medications can help relieve the symptoms and hasten the recovery. If the infection is also in the urine, it can be tested to confirm antibiotic effectiveness. Otherwise, clinical judgment is based on age and sexual exposure (discussed further in Chapter 12).

Testicular Torsion

Some testicles can be a little more mobile than others. If one becomes twisted, it can choke off the blood supply and cause the testicle to die. This is a true surgical emergency, requiring the urologist to surgically untwist the spermatic cord and suture the testicle to the inside of the scrotal wall in order to prevent repeat twisting. The other (seemingly normal) testicle will also be sutured in place to prevent it from twisting in the future.

This condition is most common during the early teen years and is very rare much beyond age thirty.[2] Testicular torsion should be a consideration in any man with sudden pain and swelling of the testicle. The diagnosis is easily confirmed or eliminated with an ultrasound that can detect the presence or absence of blood flow.

On occasion, testicular torsion can be intermittent. In these cases, the diagnosis is difficult to prove. If the story is convincing, both testicles should be sutured to the inside layer of the scrotal skin to prevent future torsion.

Recovery from this surgery is fairly quick. If performed within four hours of onset, very little damage will occur. Within six to eight weeks, you will not be able to notice any effects from the surgery. If left untreated, removal of the testicle is often the best option.

Undescended Testicles

An undescended testicle, also called cryptorchidism, occurs when the testicle fails to fall from the abdomen to the scrotum while in utero or during the first year of life. Any man with a history of an undescended testicle is at higher risk for cancer in either testicle—even the normal one.[3] Therefore, regular testicular self-exam becomes even more important. Cryptorchidism also increases the risk of future fertility issues.

An undescended testicle is usually addressed during a child's first year of life. The testicle can be surgically pulled down into the scrotum. Although this surgery will potentially improve fertility, it will not reduce the already higher incidence of testicular cancer. When detected later in life, removal is usually best because the function becomes minimal and the risk of cancer is higher.

Testicular Cancer

Although testicular cancer is relatively rare, regular testicular self-exams are recommended for early detection. The cancer is characterized by a painless lump (unlike torsion or orchitis) on or in the testicle. If the lump is separately moveable from the testicle—such as being attached to the epididymis—it is most likely not a cancer. The diagnosis can be confirmed with a scrotal ultrasound.

Testicular self-exam should be performed monthly while taking a warm shower so that the scrotal skin is loose and thin. Systematically allow *one* of the testicles to roll through your fingers using both hands for the best control. Learn which part of the scrotal contents is the epididymis so that you can recognize what belongs. Perform the same on the other side. A lump or hard spot that is part of the testicle should be brought to attention of a urologist. If the lump is clearly separate from the testicle—such as part of the epididymis—it may be safe for you just to monitor it. When in doubt, have the doctor check it out.

The initial treatment for a suspicious testicular mass is removal through a groin (lower abdominal) incision. The surgical incision is never performed on the scrotum since this

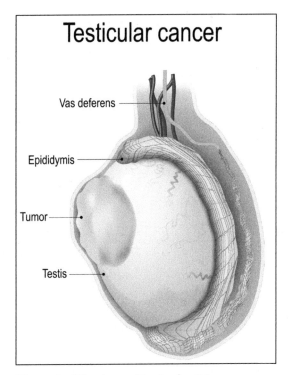

Testicular cancer

Vas deferens

Epididymis

Tumor

Testis

Figure 5: Worrisome Testicular Mass

could cause unpredictable spread of the cancer. Some special testicular cancer blood tests will be drawn prior to the surgery (alpha-fetoprotein, AFP; beta human chorionic gonadotropin, beta-hCG; and lactate dehydrogenase, LDH). A CT scan of the chest, abdomen, and pelvis will help determine if there is any spread requiring chemotherapy, radiation, or further surgery. Sometimes in the absence of spread, these treatments are offered as preventive measures in higher risk patients.

Even when testicular cancer spreads to distant areas in the body, the cure rates with chemotherapy are in the high nineties. However, catching it early on self-exam will minimize the treatment and maximize the cure.

Boxers Versus Briefs

No scrotal chapter would be complete without discussing the never-ending controversy of boxers versus briefs. The answer is that there is no answer. However, the debate remains as heated as some across-the-aisle political discussions. Briefs tend to keep the genitals supported and out of harm's way. Boxers are more breathable. Boxers allow the scrotum to regulate its own temperature which some say may lead to healthier sperm counts. However, neither style influences testosterone production or sexual prowess—unless your partner prefers one look over the other.

So, What Happened to Pomeroy?

Pomeroy saw the urologist, who diagnosed him with a spermatocele. The doctor did not think there was any infection and just prescribed over-the-counter anti-inflammatory medication. Pomeroy's symptoms resolved in a few days, although the cyst remained unchanged in size for over one year. Most importantly, this "extra part" was clearly separate from the testicle, essentially eliminating the possibility of cancer. Nevertheless, Pomeroy continues to perform regular testicular self-exams.

Bottom Line

You have a lot of important stuff hanging around down there, and a lot can go wrong. If something doesn't feel or look right, see your doctor right away. Whether it's a source of pain, infection, or a possible cancer, waiting and wondering can make it much more difficult to treat. An early detection is always better than a later regret.

References

1. Campbell, Meredith F., Alan J. Wein, Louis R. Kavoussi, Alan W. Partin, Craig A. Peters, and Patrick C. Walsh. *Campbell-Walsh Urology*. Philadelphia: Elsevier, 2016.
2. Slaughenhoupt, Bruce, William Gans, Seth Cohen, Elizabeth Takacs, Brendan Kiely Wallace, Kathleen Kieran, and Marisa Clifton. "Medical Student Curriculum: Acute Scrotum." American Urological Association, 2012. Last modified November 2022. https://www.auanet.org/meetings-and-education/for-medical -students/medical-students-curriculum/acute-scrotum.
3. Akre, Olof, Andreas Pettersson, and Lorenzo Richiardi. "Risk of Contralateral Testicular Cancer Among Men with Unilaterally Undescended Testis: A Meta-Analysis." *International Journal of Cancer* 124, no. 3 (February 2009): 687–89. https://doi.org /10.1002/ijc.23936.

CHAPTER 9

Osteoporosis in Men

Enrique, sixty-six years old and retired from accounting, is now enjoying his two favorite pastimes, golf and gardening. He has been noticing a little back discomfort in his golf swing but chalked it up to getting older. Recently, he was lifting a heavy flower pot when he felt a distinctly different and significant pain in his lower back that dropped him to his knees. The following day, an interventional radiologist injected his vertebra with a cement-like substance that corrected the fracture and the pain. Subsequently, his primary care physician obtained a bone density scan and found that Enrique had lost significant bone density. She placed him on osteoporosis treatment.

Are Men Getting Soft?

Osteoporosis is a thinning or softening of the bones. The word *osteoporosis* literally means "porous bones," and is often considered a women's disease because it is more common among women and tends to occur at a younger age in females. However, we now recognize that osteoporosis is a serious threat for men as well.

In the US, 2.2 million men over age fifty have osteoporosis and 16.9 million more have low bone mass.[1] What's more, the risk of fractures in men is 27 percent higher than the risk of prostate cancer, and men are more likely to die after a hip fracture than their female counterparts.[2]

Osteoporosis is often called a silent epidemic because there are no symptoms. Most people don't realize that they suffer from the disorder until a fracture occurs.

The Blame Game

Bone is made up of collagen, a fibrous protein that forms the structure of bone, along with minerals such as calcium. Our bones are living tissues that are constantly changing. The body continuously removes old bone and replaces it with new bone material. During childhood and our early adult years, the skeleton grows in size and strength. The amount of new bone produced by men is significantly more than produced by women. Adults reach their maximum bone density between the ages of twenty and thirty.[3] After we reach the peak bone mass, the removal of bone exceeds the formation. As a result, bones slowly become thinner, weaker, and more likely to break. If the bones lose enough mass, the patient develops osteoporosis. Certain risks can speed up the loss of bone mass.

Risk Factors in a Man's Life

There are two types of osteoporosis: primary and secondary. Primary osteoporosis is caused by bone loss related to aging or to an unknown cause. Most men with osteoporosis have at least one risk factor or secondary cause. Risk factors for osteoporosis are listed in Table 1. The most common causes of secondary osteoporosis in men include hypogonadism (low testosterone levels), alcohol abuse, smoking, use of glucocorticoid (steroid) and other medications (Table 2), gastrointestinal disorders and poor nutrition, immobilization, and too much calcium in the urine (hypercalcemia).

Table 1
Risk Factors for Osteoporosis in Men

Alcohol abuse	Chronic	HIV
Physical inactivity	obstructive	Celiac disease
Low body weight	pulmonary	Parathyroid
Calcium or	disease	disease
vitamin D	Asthma	Thyroid disease
deficiency	Cystic Fibrosis	Inflammatory
Smoking	Hypercalciuria	bowel disease
Low testosterone	Rheumatoid	(Crohn's
levels	arthritis	disease,
Medications	Chronic kidney or	Ulcerative
(see Table 2)	liver disease	colitis)
Androgen	Diabetes Mellitus	Transplants
deprivation	Cushing's	Cancer
therapy (for	syndrome	
prostate cancer)	Family history	

Hormone Levels

Having abnormally low levels of sex hormones such as testosterone, known as hypogonadism, has been linked with osteoporosis. It is normal for testosterone levels to decrease as a man ages, but testosterone levels should not drop suddenly (as estrogen levels do in women at menopause). Some medications, such as antiandrogens used in the treatment of prostate cancer,

Table 2
Medications that can cause osteoporosis

Glucocorticoids (used for inflammation)
Prednisone
Hydrocortisone
Methylprednisolone
Others
Anticonvulsants (used for seizures)
Phenytoin (Dilantin)
Phenobarbital
Carbamazepine (Tegretol and other brand names)
Primidone (Mysoline)
Anticoagulants (Blood thinners)
Heparin
Antidepressants/mood stabilizers
Fluoxetine (Prozac)
Paroxetine (Paxil)
Sertraline (Zoloft)
Lithium
Others
Cancer Chemotherapy Drugs:
Anastrozole (Arimidex)
Exemestane (Aromasin)
Letrozole (Femara)
Methotrexate
Diabetes drugs known as thiazolidinediones (TZDs)

Rosiglitazone (Avandia)
Pioglitazone (Actos)
Proton Pump Inhibitors (used for acid reflux; heartburn)
Omeprazole (Prilosec)
Pantoprazole (Protonix)
Esomeprazole (Nexium)
Lansoprazole (Prevacid)
Loop Diuretics (used for high blood pressure, heart failure)
Furosemide (Lasix)
Torsemide (Demadex)
Bumetanide (Bumex)
Thyroid hormones in excess
Levothyroxine (Synthroid)
Armour Thyroid
Aluminum-Containing Antacids
Gaviscon
Maalox
Mylanta
Immunosuppressants (used for transplant patients to prevent rejection)
Cyclosporine A (Sandimmune and other brand names)
Tacrolimus (Prograf and other brand names)

can greatly reduce testosterone levels. (See Chapter 3 on prostate cancer.) Androgen deprivation therapy (ADT) is associated with a 13 percent loss in bone density for each year of treatment and a fracture rate that is almost four times higher.[4] Interestingly, the evidence shows that low levels of estrogen in men also contributes to osteoporosis.

Alcohol Abuse

Alcohol use can affect bone mass. A recent study showed that even one or two drinks daily increases the risk of osteoporosis by 34 percent, and more than two drinks daily increases the risk by 63 percent.[5]

Alcohol affects hormones such as parathyroid hormone (PTH), calcitonin, cortisol, and growth hormones that are involved in bone growth and remodeling. Heavy or chronic drinking also causes low levels of activated vitamin D which decreases the absorption of calcium from the GI tract. Finally, alcohol abuse can lower levels of testosterone, leading to hypogonadism as discussed above. It's worth mentioning that excessive alcohol intake can also predispose people to falls.

Smoking

Like alcohol abuse, smoking has been shown to be a cause of osteoporosis in both men and women. Nicotine and other toxic substances in cigarette smoke reduce levels of PTH, testosterone, and estrogen, and increase levels of cortisol. Smoking reduces the level of vitamin D in the body and decreases the absorption of calcium. Cigarette smoke increases free radicals (destructive cells that cause damage and disease) and oxidative stress in the body and reduces the blood supply to bones (a double whammy!) It also reduces the number of bone forming cells (osteoblasts) so that less bone is made. Finally, people who smoke tend to weigh less, be less physically active, and to have poor nutrition (all of which contribute to osteoporosis).

Use of Glucocorticoids

The most common cause of secondary osteoporosis in men is the use of glucocorticoid medications. These drugs are also known as steroids and are used for the treatment of inflammatory disorders such as asthma, chronic obstructive pulmonary disease (COPD), and rheumatoid arthritis. Glucocorticoid drugs are not the same as anabolic steroids, those male hormones used by athletes to increase muscle mass. Common medications in this class are prednisone, cortisone, and methylprednisolone. These drugs have a direct effect on bone absorption and breakdown and may reduce absorption of calcium and levels of testosterone.

Bone mass decreases quickly with ongoing use of glucocorticoids. Men taking these drugs for more than three months should talk to their doctor about having a bone mineral density (BMD) test and getting testosterone levels checked regularly.

Gastrointestinal Disorders and Poor Nutrition

Several nutrients are essential for bone growth and maintenance, including calcium, magnesium, protein (amino acids), phosphorus, and vitamins D and K. People suffering from disorders of the gastrointestinal tract such as celiac disease and inflammatory bowel disease (Crohn's disease and Ulcerative colitis) as well as those who have had bariatric (weight loss), or GI surgery may have impaired nutrient absorption. In addition, those with poor diet or anorexia may be at risk for osteoporosis.

Hypercalciuria

Hypercalciuria is a disorder characterized by increased excretion of calcium in the urine. Known causes of hypercalciuria include high levels of parathyroid hormone, sarcoidosis, cancer, Cushing's syndrome, hyperthyroidism, Paget's disease, and a type of kidney disease called renal tubular acidosis. Sometimes the cause of increased calcium excretion is not known. When excessive amounts of calcium are lost in the urine, bone health may be affected.

Immobilization

Weight-bearing exercise is important for maintenance of healthy bones. When people are not able to move due to surgery, bone density declines rapidly, and disorders like Parkinson's disease, multiple sclerosis, stroke, or spinal cord injury may occur. It is crucial for people to resume weight-bearing exercise, like walking, as soon as possible after a period of immobility.

Screening and Diagnosis

A physical exam and careful medical history are important for the detecting of osteoporosis. Your physician may also want to obtain blood and urine tests. Recommended guidelines for osteoporosis screening in men vary by the organization. The National Osteoporosis Foundation (NOF) and the Endocrine Society recommend testing bone mineral density (BMD) in all men who are seventy years of age or older. In addition, men between the ages of fifty and sixty-nine should be tested if they have risk factors for fractures or have suffered a fracture. The test that is considered the gold standard for determining bone density is called the dual energy x-ray absorptiometry test

(DEXA). The DEXA test measures bone mineral density at the hip and lower spine using a "T-Score." A T-score of -2.5 or less is indicative of osteoporosis. A score greater than -1 is normal. A T-score between -1 and -2.5 shows that bones are becoming weak. This is called osteopenia. Bone density tests are quick, painless, and involve very low amounts of radiation. In fact, you are exposed to more radiation on a cross-country flight. DEXA testing should be repeated every one to two years.

Osteoporosis and Fracture Prevention

Wouldn't it be better to prevent osteoporosis than to be subject to a painful fracture? If you are a man at risk for osteoporosis, there are some simple steps that you can take to prevent the disease from occurring.

Diet—It's Not Only About Calcium

The saying "You are what you eat" comes into play when we talk about the prevention of osteoporosis. A balanced diet with plenty of protein and calories is important when it comes to overall health as well as for osteoporosis prevention. Of course, optimal amounts of calcium, vitamin D, and magnesium are also essential for maintenance of bone density. The Food and Nutrition Board (FNB) at the Institute of Medicine of the National Academies recommends that adult men consume a total of 1,000 mg per day of calcium up until the age of seventy. Men over seventy should have a total calcium intake of 1,200 mg per day. Dairy products such as milk, cheese, and yogurt are good sources of calcium as well as leafy green vegetables such as broccoli and kale and bony fish like salmon. Many foods are fortified with calcium. See Table 3 for calcium content of selected food products. Those who don't obtain enough calcium in their diet should take calcium supplements to ensure adequate intake. Vitamin D is also important for bone health and prevention of osteoporosis. Dietary vitamin D should be 600 international units per day up to age seventy and 800 IU daily for those over the age of seventy. Good sources of vitamin D include milk, egg yolks, and fatty fish. Men with low levels of vitamin D should receive supplements. Magnesium is also important for bone mineralization as well as other critical functions in the body. It is found in legumes, nuts, grains, and fish. Consult your health care professional before taking any nutritional supplements for osteoporosis prevention.

Table 3: Calcium Content of Selected Food Items		
Food	Amount	Calcium Content
Milk	1 cup	300mg
Cottage cheese	½ cup	65mg
Ice cream	½ cup	100mg
Sour cream	1 cup	250mg
Soy milk	1 cup	300-400mg
Yogurt	1 cup	450mg
Hard cheese	1 oz	200mg
Parmesan cheese	1 tbsp	70mg
Swiss cheese	1 oz	270mg
Broccoli (cooked)	1 cup	180mg
Kale (raw)	1 cup	55mg
Spinach (cooked)	1 cup	240mg
Figs (dried)	1 cup	300mg
Orange juice (fortified)	8 oz	300mg
Almonds	1 oz	80mg
Canned salmon	3 oz	180mg
Canned sardines	4 oz	350mg

Exercise

Prevention of osteoporosis offers us another incentive to get up from the couch. Weight bearing exercises such as running, walking, and dancing stimulate osteoblasts to strengthen our bones. Swimming is an excellent aerobic exercise, but since it isn't weight bearing, it won't help to prevent osteoporosis. People who have not been active in the past or who are older might want to begin slowly, perhaps with walking. Regular exercise is recommended—thirty minutes of physical activity at least five times weekly is ideal. Weight training is especially important because it not only assists in strengthening our bones, but it also increases muscle tone and helps with coordination and balance, which may help to prevent falls.

Falls

Falls can happen at any age and can significantly increase the risk of fractures. It's important to reduce our risk of falls by taking the following steps:

- Wear comfortable shoes that fit properly with non-skid soles; make sure laces are tied.
- Avoid walking on slippery surfaces such as snow, ice, and wet floors.
- Remove throw rugs or make sure they don't slide by using double-sided tape or slip-resistant pads.
- Place electrical cords out of walkways.
- Walk in areas that are well-lit inside and outside the home.
- Keep a night-light between your bed and the bathroom.
- Use handrails and grab bars in stairwells and bathrooms.
- Make sure to have vision checked regularly.

Lifestyle Choices

To prevent osteoporosis, it is important to quit smoking and avoid having more than two drinks daily. Excess drinking may also cause falls.

Treatment of Osteoporosis

The decision to administer pharmacologic treatment for osteoporosis is based on clinical evaluation, risk of fractures, and bone marrow density measurements. In 2008, the World Health Organization (WHO) developed a tool to assess risk of fractures. This tool is called the fracture risk assessment tool (FRAX). It can be accessed at http://www.shef.ac.uk/FRAX. The fracture risk assessment tool estimates the ten-year risk of major osteopathic fracture and hip fracture by factoring in risk factors like age, weight, sex, history of fracture, smoking, alcohol use, and bone marrow density. Guidelines recommend treatment for men who are fifty years of age or older and have one of the following scenarios:

- Have suffered a fracture in the hip or spine;
- Have a T-score ≤-2.5 at the femoral neck (top of leg bone) or spine with no reversible cause;
- Have low bone mass (T-score between -1.0 and -2.5) at the femoral neck or spine) and a 10-year risk of hip fracture ≥3 percent or a 10-year risk of major fracture ≥20 percent.

Men who are above the age of fifty who are taking glucocorticoids should also be considered for treatment.

Medications for Osteoporosis in Men

Several types of medications are FDA approved for treatment of osteoporosis in men. These include the bisphosphonates, parathyroid hormones, a monoclonal antibody, and testosterone. (Table 4)

Table 4	
Medications for Osteoporosis in Men	
Bisphosphonates	Teriparatide (Forteo)
Alendronate (Fosamax)	Monoclonal Antibody
Risedronate (Actonel, Atelvia)	Denosumab (Prolia)
Zoledronic acid (Reclast)	Testosterone
Parathyroid hormone	

The Bisphosphonates

These medications include Alendronate (Fosamax), Risedronate (Actonel or Atelvia), Ibandronate (Boniva), and the injectable Zoledronic acid (Reclast). These medications work by inhibiting the osteoclasts that break down and remove bone. They are the first line for treatment of osteoporosis in men.

You should take Fosamax, Actonel, and Boniva first thing in the morning on an empty stomach with a full eight-ounce glass of water. You should take Atelvia just after breakfast with four ounces of water. Binosto is an effervescent tablet that should be placed into four ounces of water. Wait five minutes after it dissolves and stir before drinking. After taking these medications, wait at least thirty minutes before eating or taking other medications or vitamins, calcium, or antacids. Don't lie down or recline for at least thirty minutes after taking these bisphosphonates. Reclast is given by your health care professional into your vein as an IV medication. The infusion will last at least fifteen minutes. It is usually given once a year for treatment of osteoporosis. Patients may eat before receiving Reclast and should drink plenty of fluids throughout the day. Some people who take Reclast experience flu-like symptoms within one to three days after receiving the first dose. The symptoms may include fever and joint/muscle pain. Taking acetaminophen usually lessens discomfort.

Most people who take bisphosphonates don't experience any serious side effects. Common side effects include nausea, abdominal pain, indigestion, heartburn (dyspepsia), joint and/ or muscle pain, and diarrhea or constipation. Rare side effects of the bisphosphonates include inflammation or ulcers of the esophagus and osteonecrosis of the jaw. Osteonecrosis of the jaw occurs when the jawbone doesn't receive an adequate blood supply and thereby is also starved of nutrients. It's important for those taking bisphosphonates to have good oral health— brush and floss teeth daily and see a dentist on a regular basis.

Parathyroid Hormone

Teriparatide (Forteo) is a synthetic version of parathyroid hormone (which is naturally found in the body and regulates bone growth). Forteo is the only medication that stimulates the formation of bone. It is given as a daily injection under the skin for two years. Forteo is a treatment for men with severe osteoporosis or for those whose osteoporosis has not improved with bisphosphonates. Side effects associated with Forteo are mild and include pain at the injection site, dizziness, nausea, and temporary high calcium levels four to six hours after the dose. The use of Forteo is limited to two years due to the possible risk of bone cancer if large doses are given. Patients should rotate injection sites in the thigh or abdominal areas and sit when injecting the medication.

Denosumab (Prolia)

Denosumab (Prolia) is a type of drug known as a monoclonal antibody. Monoclonal antibodies are proteins that bind to a type of cell in the body. Denosumab (Prolia) binds to the osteoclast cells in the body to prevent these cells from breaking down bone. Prolia is a treatment for men with osteoporosis who are at high risk for fractures, including those who have suffered from a previous fracture. It is also used for patients who have failed or cannot take other treatments for osteoporosis. It is given as an injection under the skin every six months. Prolia is generally well tolerated by patients, but some side effects can occur such as skin infections and rash. Like Forteo, Prolia can cause a mild, temporary elevation in calcium levels.

Testosterone

According to the latest Endocrine Society guidelines, treatment with testosterone should be given to men with low levels of the male hormone and high risk of fracture. The society also recommends that these men also receive a bisphosphonate or Teriparatide along with testosterone therapy. Studies have shown that testosterone increases bone marrow density in men with low levels of testosterone. So far, studies have not assessed the effects of combination therapy of testosterone with bisphosphonates or other drugs for osteoporosis.

So, Whatever Happened to Enrique?

Enrique's physician prescribed calcium, vitamin D, and magnesium supplements along with a once-weekly Fosamax tablet, Enrique limits his drinking, mostly to a little red wine from time to time. He recovered from his vertebral fracture and his bone density has greatly improved. He continues to enjoy his favorite retirement activities, golfing and gardening.

The Bottom Line

Osteoporosis is starting to be recognized as a significant health problem for men, but men are still less likely to be diagnosed and treated than females. Men can take steps to prevent bone loss before it happens by eating well, taking calcium and vitamin D supplements, avoiding unhealthy habits like excessive alcohol consumption and smoking, and exercising on a regular basis. Fall prevention strategies can also be an effective way to avoid fractures. Make sure to see your physician on a regular basis and to ask about your risk for osteoporosis. Those at risk can take steps to avoid further loss of bone mass and fractures. For those men who are diagnosed with osteoporosis, effective treatments are available.

References

1. Sarafrazi, Neda. "Osteoporosis or Low Bone Mass in Older Adults: United States, 2017–2018." Interview by *NCHStats* (blog). National Center for Health Statistics, accessed April 24, 2022. https://nchstats.com/2021/03/31/osteoporosis-or-low-bone-mass-in-older-adults-united-states-2017-2018.
2. International Osteoporosis Foundation. "Epidemiology of Osteoporosis and Fragility Fractures." Accessed April 24, 2022. https://www.osteoporosis.foundation/facts-statistics/epidemiology-of-osteoporosis-and-fragility-fractures.
3. National Institute of Arthritis and Musculoskeletal and Skin Diseases. "Bone Health and Osteoporosis." National Institutes of Health, accessed April 24, 2022. https://www.bones.nih.gov/health-info/bone/osteoporosis/men.
4. Lee, Daniel, and Shannon Gallagher. "Aqua in Action: Baseline Bone Density Evaluation among Men on Androgen Deprivation

Therapy." American Urological Association (website), March 1, 2022. https://www.auanews.net/issues/articles/2022/march-2022/aqua-in-action-baseline-bone-density-evaluation-among-men-on-androgen-deprivation-therapy.

5. Cheraghi, Zahra, Amin Doosti-Irani, Amir Almasi-Hashiani, Vali Baigi, Nasrin Mansournia, Mahyar Etminan, and Mohammad Ali Mansournia. "The Effect of Alcohol on Osteoporosis: A Systematic Review and Meta-Analysis." *Drug and Alcohol Dependence* 197, no. 1 (April 2019): 197–202. https://doi.org/10.1016/j.drugalcdep.2019.01.025.

CHAPTER 10

Chronic Pelvic Pain

Vladimir (Vlad), age forty-three, has pain and discomfort in the area under his scrotum. He goes to the restroom more than twenty times a day. He describes the toilet as his new best friend! He gets up five to seven times a night just to urinate. He is worn out and tired upon awakening in the morning and believes that his symptoms are affecting his work and his relationship with his wife. His urine examination is negative, with no evidence of blood in his urine or evidence of a urinary tract infection. He has pain and discomfort with urination. He was treated with multiple medications including antibiotics and muscle relaxants without relief. Vlad and his wife are not happy campers.

Chronic Pelvic Pain

Chronic pelvic pain (CPP) affects both women and men. Though CPP has traditionally been considered a women's disease, men also get pelvic pain. For both sexes, many of the challenges are similar, including exually related pain.

Men with CPP face challenges of their own. The diagnosis is often missed because CPP is less common in men, and its symptoms overlap with those of other more common conditions in men. CPP has multiple other names: interstitial cystitis; chronic non-bacterial prostatitis; bladder pain syndrome; and pelvic floor dysfunction.

For the remainder of this chapter, all diagnoses in men with chronic pelvic pain will be referred to as CPP. In medicine, any condition that has many different names indicates some confusion in the medical community, and this means there is always a challenge for both patients and clinicians.

The symptoms for CPP include pelvic pain or a dull ache in the pelvic area, urinary urgency, painful urination, pain with intercourse, and urinary frequency. CPP can affect a man's social life, sleep, and even his ability to function effectively in the workplace. Most men with this condition will have quite frequent urination, and there are some at the extremes who might have to urinate forty to sixty times a day. It is no wonder that a CPP patient's best friend is the nearest toilet. Men with CPP often state that they are prisoners of their pelvis. Men with CPP awaken multiple times a night because they feel as if they have a full bladder, along with pelvic pain.

But these symptoms overlap with other conditions that are more common in men, especially chronic bacterial prostatitis, urinary tract infections, and benign enlargement of the prostate gland. (See Chapter 2.)

Today, the trend is to consider chronic pelvic pain and interstitial cystitis (IC) as the same condition. Perhaps the hallmark of CPP is the presence of urinary symptoms, pain and discomfort, along with a urine examination, a negative urine culture, i.e., no bacteria identified, and no response to antibiotics.

If a man with CPP also has chronic lower urinary tract symptoms, such as urgency, frequency, nocturia (nighttime urination), pain with bladder filling, pressure above or in front of the pubic bone, or painful urination, and does not respond to standard therapies for prostatitis, he may have CPP. Clinical experience suggests that if these patients are treated specifically for CPP, they tend to do better than if they are treated only with typical chronic prostatitis therapies, such as alpha-blockers and/or antibiotics.

It is often difficult to differentiate prostatitis from CPP. With prostatitis, the pain is experienced in the area between the testicles and anus, referred to as the perineum. Prostatitis is characteristically accompanied by urinary symptoms, and these affected men may have an abnormal digital rectal exam with a tender prostate and an abnormal urinalysis. If the cause is due to a urinary infection, the urine culture will confirm the infection.

Men are more reluctant than women to say they have pain and may be less apt to share that they have bladder and pelvic pain. You know, that macho thing! Sociological studies show that this tendency is ingrained in a man early in life. By age five or six, boys are less likely than girls to express hurt or distress. Little boys are told "Don't cry," "Be a man," and "Suck it up!" Research also shows that men's coping skills are not as well developed as women's coping skills. These tendencies mean that men may not seek pain control as soon as they need to and may need more help developing coping skills in order to be more comfortable sharing their medical concerns with a health care professional.

Causes of CPP

The bladder is shaped like a balloon and is located in the pelvis. It can expand and contract just like a balloon. It has two tubes, called ureters, which transport urine that is made in the kidneys to the bladder where it is stored. There is another tube, the urethra, located in the penis, which allows the urine to leave the bladder when the man has the desire to urinate.

Normally the bladder can expand and hold fluid or urine without any discomfort. However, in patients with CPP, the expansion stretches the muscles, which irritates the nerves that supply the bladder and causes annoying discomfort all the way to severe pain. As a result, men with CPP have pain as the bladder expands, which also makes them feel that they always have to go to the bathroom. Most men find that the pain in the pelvis will subside after they urinate. In severe cases, even after urinating, the pain will persist.

The best explanation we have of the cause of CPP is that there is a defect in the lining of the bladder that enables a toxic chemical in the urine to penetrate and irritate the muscles and the nerves that supply the bladder, resulting in the urge to urinate. As the lining becomes leaky, scar tissue forms in the muscles of the bladder, which decreases the capacity of the bladder to expand when fluid is added to the bladder from the kidneys. This situation then causes men to have a smaller bladder capacity and to have to urinate more frequently.

No one knows what causes chronic pelvic pain (CPP), or interstitial cystitis/painful bladder syndrome (IC/PBS), but doctors believe that it is a real physical problem and not a result, symptom, or sign of an emotional problem. Because the symptoms of CPP are varied, most researchers believe that it represents a spectrum of disorders rather than one single disease.

One area of research on the cause of CPP has focused on the layer that coats the lining of the bladder called the glycocalyx, made up primarily of substances called mucins and glycosaminoglycans (GAGs). This layer normally protects the bladder wall from any toxic contents in urine. Researchers have found that this protective layer of the bladder is "leaky" in about 70 percent of CPP patients and have hypothesized that this may allow substances in urine to pass into the bladder wall where they might trigger CPP directly, or may make these patients susceptible to other chemicals in the urine, including those from foods or beverages.

Along with altered permeability of the bladder wall, researchers are also examining the possibility that CPP results from decreased levels of protective substances in the bladder wall. Reduced levels of GAGs (discussed previously) or other protective proteins might also be responsible for the damage to the bladder wall seen in CPP.

No matter what the mechanism for disruption of the bladder lining, potassium is one substance that may be involved in damage to the bladder wall. Potassium is present in high concentrations in urine and is normally not toxic to the bladder lining. However, if the tissues lining the inside of the bladder (urothelium) are disrupted or are abnormally leaky, potassium could then penetrate the lining tissue and enter the muscle layers of the bladder where it can cause damage and promote inflammation and pain.

Researchers have isolated a substance known as anti-proliferative factor (APF) that appears to block the normal growth of cells that make up the lining of the bladder. APF has been identified almost exclusively in the urine of people suffering with CPP. Research is under way to clarify the potential role of APF in the development of CPP.

Increased activation of sensory nerves (neurologic hyper-sensitivity) in the bladder wall is also thought to contribute to the symptoms of CPP. In addition, cells known as mast cells within the bladder wall, which play a role in the body's inflammatory response to injury, release chemicals that are believed to con-tribute to the symptoms of CPP.

Other theories about the cause of CPP are that it is a form of autoimmune disorder (in which the body's own immune system attacks the body) or that infection with an unidentified organism may be producing the damage to the bladder and the accompanying symptoms.

Diagnosis of CPP

It is very important for male patients to have a thorough diagnostic workup. This includes a careful history and a physical exam, including a digital rectal exam to check the prostate gland. Urine tests are ordered and include a urinalysis and a urine culture to be certain that there is no blood in the urine and no urinary tract infections. Another test which is commonly performed is a cystoscopy—a look into the bladder with a lighted tube. In approximately 5 percent of men with CPP, there are characteristic ulcers, referred to as Hunner's ulcers, located in the inner lining of the bladder, and revealed through a cystoscopy.

Another diagnostic test is hydrodistention of the bladder, or filling the bladder with sterile water to determine the bladder capacity. Hydrodistention is performed under general or regional anesthesia. This workup will help rule out other medical conditions that cause pelvic pain and will help to confirm chronic pelvic pain as a diagnosis.

Urinalysis and Urine Culture

These tests can detect and identify the most common bacteria in the urine that may be causing IC/PBS-like symptoms. A urine sample is obtained either by catheterization with a small tube inserted into the penis, or more commonly by the "clean catch" method. For a clean catch, the man washes his genitals before collecting a sample of urine "midstream" in a sterile container. White and red blood cells and bacteria in the urine suggest an infection of the urinary tract that can be treated with antibiotics. If urine is sterile for weeks or months while symptoms persist, a doctor may consider a diagnosis of CPP.

Culture of Prostatic Secretions

Using a digital rectal examination, the doctor can obtain a sample of prostatic fluid. This fluid is examined under the microscope for signs of an infection such as red and white blood cells, and also can be cultured for bacteria. Prostatic infections can be treated with antibiotics.

Potassium Sensitivity Test

The intravesical potassium sensitivity test (PST) has been developed to evaluate the leakiness of the protective lining of the bladder. Some experts recommend its use in the evaluation of CPP, but we feel it is antiquated, and the pain and the discomfort are a significant deterrent for most doctors trying to diagnose CPP. At the present time, we are not recommending the use of the PST.

Lidocaine Instillation

Filling the bladder with a solution containing the local anesthetic drug, lidocaine, has been described as an "anesthetic bladder challenge." Improvement of symptoms after lidocaine has been instilled into the bladder suggests CPP. However, this test is not specific for CPP and is not routinely performed.

The Guidance Test

At times a man can actually have an infection that is very difficult to detect; he may even have a negative culture. The Guidance Test actually looks for the DNA traces from bacteria in the urine or prostate secretions. This test also allows the physician to know what antibiotics may be effective. In situations where it really "seems" like an infection, based upon symptoms, but with a negative culture, this test can shed light on what may be going on.

Cystoscopy Under Anesthesia with Bladder Distension

During cystoscopy, the urologist uses a cystoscope—an instrument made of a hollow tube about the diameter of a drinking straw with several lenses and a source of light—to look inside the bladder and urethra. The urologist will also distend or stretch the bladder to its capacity by filling it with a liquid or inert gas such as carbon dioxide. Because bladder distension is painful, the patient is often given either regional or general anesthesia before the doctor inserts the cystoscope through the urethra into the bladder. Cystoscopy with distension of the bladder with sterile water can detect inflammation (visually or with biopsies), a thick and stiff bladder wall, and Hunner's ulcers. After the fluid has been drained from the bladder, small red spots, called glomerulations, looking a bit like measles, indicate enlarged blood vessels that pinpoint areas of bleeding in the bladder lining.

The cystoscopy also allows measurement of a patient's bladder capacity—the maximum amount of liquid or gas the bladder can hold under anesthesia (without anesthesia, capacity is limited by either pain or a severe urge to urinate.) Most people without CPP have normal 10-15 ounces of bladder capacity under anesthesia. A smaller bladder capacity due to scarring of the bladder wall helps to support the diagnosis of CPP.

Cystoscopy is recommended only to exclude other possible causes of symptoms and not as the definitive diagnostic test for CPP.

It is important to note that the distension often performed with cystoscopy may lead to relief of symptoms in some patients with CPP; this generally lasts from several weeks to months following the procedure. If the patient does get relief for a time, he may want to repeat the procedure if symptoms recur.

Bladder Biopsy

A biopsy is a microscopic examination of a small sample of tissue. Samples of the bladder and urethra may be removed during cystoscopy and examined under a microscope by a pathologist. A biopsy helps to exclude bladder cancer and also may confirm the presence of mast cells or inflammation of the bladder wall that is consistent with a diagnosis of IC/PBS. Nevertheless, there is nothing on the biopsy that can make an absolute diagnosis of IC/PBS. Unfortunately, there are no lab tests that can distinguish CPP from nonbacterial prostatitis or urethritis.

We can summarize the diagnosis of CPP by stating that the condition is difficult to diagnose and is often considered to be a diagnosis of exclusion, meaning that other diagnoses such as infection, cancer etc. have been eliminated. Though treatments can reduce symptoms, there's no cure. In most clinical situations, diagnosing the problem with the patient's history, the clinical presentation, and a physical exam is all there is.

Treatment for the Pain Down There

Many methods have been tried to provide relief from the symptoms of CPP. However, no standard or universally accepted treatment has been found that reliably and invariably resolves this debilitating condition.

Removing the Offending Foods and Fluids from Your Diet

Many men with CPP find that eliminating or reducing their intake of potential bladder irritants may help to relieve their discomfort. The most irritating foods can be summarized as the "four Cs." The four Cs include carbonated beverages, caffeine in all forms (including chocolate), citrus products, and food containing high concentrations of vitamin C and potassium.

If you find that these things irritate your bladder, you may also wish to avoid related foods such as tomatoes, pickled foods, alcohol, and spices. Artificial sweeteners may also aggravate symptoms in some men. If you think certain foods make you feel worse, try eliminating them from your diet. Reintroduce them one at a time to determine which, if any, affect your signs and symptoms. Additionally, making your urine less acidic by taking alkalinizing agents such as citrate may improve symptoms of frequency and pain associated with CPP. Citrates, which include potassium or sodium citrate, tricitrates, and citric acid, either alone or in combination (Bicitra, Citrolith, Oracit, Polycitra, Urocit-K), are usually used to prevent certain

Table 1
Common bladder irritants to avoid

Tomatoes (includes ketchup, salsa, pizza sauce)
Caffeine
Chilies/Spicy foods
Chocolate
Citreous fruits and drinks: Oranges, limes, and lemons
Alcoholic beverages: beer, wine, liquor
Carbonated beverages: soft drinks, soda water, energy drinks
Spicy foods: salsa
Sweeteners: artificial and natural sweeteners
Processed foods
Onions
Cranberries
Vinegar
Raw onions

types of kidney stones. But because they make the urine less acidic, they may help relieve bladder pain in men with CPP. These remedies are available by prescription.

Most men can identify the culprits by conducting a simple elimination diet. Stop eating all of the foods on the bladder irritant list shown in Table 1. We recommend that you adhere to this elimination diet for *two weeks*. If the bladder symptoms improve, then you can be fairly sure that one or several of the foods or liquids was irritating your bladder. Next, begin by eating just one food from the bladder irritant list food list. If you don't develop any symptoms of pelvic pain or increased urgency or urinary frequency within twenty-four hours, then you can be reasonably certain that this is not one of the foods or fluids that is affecting your CPP. You can then add one new food or liquid each day until you have identified the offending foods that are guilty of making you so miserable.

Oral Medications for CPP

Because the cause of CPP is unknown, numerous empirical therapies have been tried. Oral medications have given relief to some men and are commonly the first-line treatment. The only FDA-approved drug for the treatment of interstitial cystitis/ painful bladder syndrome (IC/PBS) is pentosan polysulfate sodium, or Elmiron. Elmiron is a very weak blood thinner or anticoagulant. It probably works by coating the inner lining of the bladder that becomes damaged by the disease, and helps to restore the inner lining so that the bladder lining no longer leaks. Perhaps Elmiron prevents toxins in the urine from penetrating the inner lining and irritating the nerves under the bladder lining that cause the severe pain, urinary frequency and urgency. Unfortunately, Elmiron may take between three to six months to improve urinary symptoms and reduce pelvic pain.

A decrease in pain is often one of the first reported positive effects, with improvement in frequency and urgency of urinary symptoms occurring after pain relief. It appears that about 35 to 40 percent of CPP men who take Elmiron notice marked improvement. Part of the reason that Elmiron may take so long to work is that it often has to undo months or years of damage to the bladder's inner lining resulting from CPP. Elmiron itself does not heal the bladder wall; instead, it offers a layer of protection to the bladder while it heals, which is often a drawn-out process. As such, it's best to stay on Elmiron for at least a six-month trial period, as it may take that long to see any positive effects.

Elmiron is a reasonably safe drug, and only a few men with CPP experience side effects such as diarrhea or nausea. A very rare side effect is hair loss, which occurs in about 3 percent of the men who use Elmiron. Fortunately, the hair loss is temporary and the hair will grow back once the medication is discontinued. Recent reports claim that Elmiron may cause damage to the macula of the eye. Users might be advised to visit with an ophthalmologist for a full eye exam.

Other oral medications include anticholinergic drugs such as Ditropan, Detrol, Vesicare, Toviaz, Enablex, and Sanctura, which are used to decrease bladder spasms. Another class of drugs called beta-3 adrenergic agonists include Myrbetriq and Gemtesa; they can also relieve frequency and painful spasms. You may receive bladder analgesics such as pyridium, which turns the urine red, or Prosed, which turns the urine blue. These medications allow you to urinate in living color! Both of these drugs can temporarily relieve the pain or burning associated with urination. Over-the-counter forms of phenazopyridine hydrochloride (Azo-Standard, Prodium, and Uristat) may provide some relief from urinary pain, urgency, frequency, and burning. Higher doses of the phenazopyridine (Pyridium) are available with a doctor's prescription.

Other medications that may reduce the symptoms of CPP include sedatives for those men who have trouble falling and staying asleep. Antidepressants have also been effective. Some of the most commonly used sleep aids include tricyclics such as amitriptyline (Elavil), doxepin (Sinequan) and imipramine (Tofranil), given at bedtime. These agents may be started at very low dosages and gradually increased until symptom relief is obtained or the side effects become bothersome. Use of hydroxyzine (Atarax), an antihistamine, is also effective especially if the biopsy of the bladder shows an excess of mast cells, which release the histamine responsible for the pain in the bladder.

Other oral medications that may be used to treat CPP include low doses of antidepressants of the tricyclic group such as amitryptiline (Elavil). It is believed that tricyclic antidepressants can help reduce the hyperactivation of nerves within the bladder wall. The anti-seizure medication gabapentin (Neurontin, Gabarone, Gralise, Horizant, FusePaq Fanatrex) also has been used to treat nerve-related pain as well as the pain of CPP. Oral antihistamines, such as hydroxyzine, also may be prescribed to help reduce allergic symptoms that may be worsening the patient's CPP symptoms.

Leukotrienes, which are substances produced by some immune system cells and mast cells, promote inflammation. Drugs that block leukotrienes are new and are being used in the treatment of asthma and allergy. They include the prescription medicines:

- Montelukast (Singulair)
- Zafirlukast (Accolate)
- Zileuton (Zyflo)

New evidence has implicated leukotrienes in inflammation of the bladder in men with CPP. The leukotriene receptors have been found in the bladder muscle in patients with CPP. Patients who took Singulair for three months showed significant reductions in urinary frequency and pelvic pain. MSM, a supplement often used for arthritis pain, is available over-the-counter. In doses between 1000mg and 3000mg daily, it helps men with CPP. One would need to try it for at least two to three months to assess its benefits.

In many patients we use not just one or two medications, but several all taken together; each one has a different impact on the bladder.

Medications Placed into the Bladder

If you take medication by mouth, the drug must be absorbed through the gastrointestinal tract, then introduced into the bloodstream, and finally filtered by the kidney and excreted in the urine, where the drug can do its job against the abnormal lining of the bladder. However, you can perform a bloodstream-kidney bypass by inserting the medication directly into the bladder.

By placing a tiny catheter or tube through the urethra, the bladder is drained of urine and then medication is placed into the bladder where it heals the altered lining. The medication only remains in the bladder for a few minutes, and the man goes to the restroom and urinates the remaining medication into the toilet. This treatment option has almost no side effects because only a very small amount of the medication is absorbed from the bladder. Medication instilled into the bladder is usually not painful or even uncomfortable.

Many agents have been instilled into the bladder in attempts to relieve symptoms. Today, the most commonly used medication is dimethyl sulfoxide (DMSO) or RIMSO. DMSO is approved by the FDA for the treatment of CPP. Although DMSO is an anti-inflammatory agent, it seems to have some pain-relieving properties as well. As is true of other therapies for CPP, there is no way to predict which men may or may not respond to DMSO.

DMSO treatment regimens vary and have included instillation weekly for six weeks, and men especially respond to a "booster" dose every month. There is no uniformity of choice regarding treatment schedules. Most likely, each physician uses a slightly different approach; as yet, there is no "right" way to use DMSO. It is probably safe to say that if a man with CPP does not experience relief within a few treatments, further instillation of DMSO is unlikely to be effective. The most common side effect of DMSO is a complaint of a garlic-like taste and odor after instillation. Most men will find that sucking on a peppermint candy after the treatment alleviates the garlic taste.

Following the initial course of treatment, some men achieve long-term remission, but most men will eventually relapse. Additional treatment schedules for those who relapse vary but usually consist of instillation every four-to-six weeks. Those who learn how to insert the small catheter and instill the DMSO in the privacy of their homes can use the treatment whenever symptoms occur without having to make a doctor's appointment.

DMSO is sometimes combined with other medications such as heparin, steroids, bicarbonate, and a local anesthetic (xylocaine) as a bladder treatment. A bladder "cocktail" consisting of heparin, lidocaine, and sodium bicarbonate has been effective in some men. This cocktail is given several times a week for several weeks, and many CPP sufferers respond immediately to this treatment.

Botox® and Your Bladder

Botox®, which has been used for years to wipe away those wrinkles around the mouth and forehead, has been investigated for injection into the bladder muscle to reduce bladder spasms in men with CPP who have not responded to oral medication or bladder instillations. The procedure is usually done in the operating room where a very tiny needle is inserted through a cystoscope and the bladder muscle is injected at multiple locations with Botox®. Improvement is expected within two to three weeks, and if there is improvement, Botox® injections will often last from four to six months and then the procedure can be repeated.

Botox® is made from the botulism toxin. It has been modified and has been utilized in FDA-approved cosmetic procedures such as the treatment of wrinkles. By injecting Botox® into the patient's bladder, the nerves are "put to sleep," calming their harmful influence on the bladder. Unfortunately, Botox® does not cure the problem and must be repeated from time to time.

Physical Therapy

Although physical therapy (PT) is well known for helping men recover from orthopedic surgery, it also helps men with soft tissue injuries. PT has been shown to be helpful to ease muscular spasm in the pelvic floor. Although it might be difficult to think about going through this therapy because pelvic floor muscle massage requires a digital rectal exam, PT can offer tremendous relief. PT also includes learning how to avoid activities that exacerbate your symptoms by releasing muscle trigger points and using techniques that keep the pelvic muscles relaxed. You can be taught to treat yourself so you can do the exercises in the comfort and privacy of your own home. It helps to involve your spouse or significant other in PT, as he or she can help massage muscles and trigger points you cannot reach yourself. We suggest you find a pelvic floor physiotherapist who has experience in treating men with CPP. Pelvic floor physiotherapists sometimes combine the PT technique with biofeedback.

You Must Prevail When All Else Fails

The vast majority of men with CPP can be helped with medications, drugs placed in the bladder, and hydrodistension. However, there are a few unfortunate men for whom nothing seems to help. They have tried all of the conservative approaches and have seen multiple physicians and experts who treat CPP, and yet the men remain miserable and incapacitated by the disease. In these few men, a surgical option may be the only solution. Attempts can be made to sever the nerves to the bladder that appear to be associated with the pain and discomfort. Occasionally, some men will have a part or even the whole bladder removed. For those men with CPP who have a very small bladder that holds only a small volume of urine, their bladder can be effectively enlarged by attaching a piece of intestine to the bladder in order to increase capacity.

Nerve Stimulation

If you have tried diet changes, exercise, and medicines and nothing seems to help, you may wish to think about nerve stimulation. This treatment, called neuromodulation, sends mild electrical pulses to the nerves that control the bladder.

At first, you may try a system that sends the pulses through electrodes placed on your skin. If this therapy works for you, you may consider having a device put in your body. The device delivers small pulses of electricity to the nerves around the bladder.

For some patients, nerve stimulation relieves bladder pain as well as urinary frequency and urgency. The nerve stimulation treatment consists of electrical pulses that block the pain signals carried in the nerves. If your brain doesn't receive the nerve signal, you don't feel the pain. For others, the treatment relieves frequency and urgency, but not pain. For still other patients, it does not work.

Mild electrical pulses can be used to stimulate the nerves to the bladder—either through the skin or with an implanted device. The method of delivering impulses through the skin is called transcutaneous electrical nerve stimulation (TENS). With TENS, mild electric pulses enter the body for minutes to hours, two or more times a day, either through wires placed on the lower back or just above the pubic area—between the navel and the pubic hair. The electrical pulses may increase blood flow to the bladder, strengthen pelvic muscles that help control the bladder, or trigger the release of substances that block pain.

TENS is relatively inexpensive and allows people with CPP to take an active part in treatment. Within some guidelines, the patient decides when, how long, and at what intensity TENS will be used. It has been most effective in relieving pain and decreasing frequency of urination. Smokers do not respond as well as nonsmokers. If TENS is going to help, improvement is usually apparent in three to four months.

A person may consider sacral neuromodulation, having a device such as InterStim implanted to deliver regular impulses to the bladder. A wire is placed next to the tailbone and attached to a permanent stimulator under the skin. The FDA has approved this device to treat CPP when other treatments have failed.

Acupuncture

Another option being tested as an alternative treatment modality is the use of acupuncture, with two to three treatments each week for several months. The mechanism of action on the use of acupuncture for the treatment of CPP is not known. Perhaps the symptoms of frequency and urgency are alleviated because pressure in the bladder and urethra is reduced by an action of the acupuncture needles on blocking excessive nerve impulses. If the acupuncture treatment can adjust the nerve impulses to the bladder, it may be possible to alleviate some of the sense of urgency that occurs with small amounts of urine in the bladder, and it may also reduce some of the inflammatory processes that occur in the bladder lining.

Coping Tips for Patients with CPP

There is probably no problem in the pelvis that is more demanding and challenging than CPP. Emotional support is very important in coping with ongoing pain and the necessity of going to the restroom so frequently. Family and friends can supply you with ongoing support.

Remember that you are not alone. There are excellent support groups that meet on a regular basis, and you can also find support groups online that are usually very helpful. Any chronic condition, especially one that causes pain, can make an individual feel alone and isolated. It weighs heavily on your mind and spirit and desire to carry on with a normal life.

A personal support system can be the critical factor in allowing such a patient to continue to enjoy a quality of life and to continue to work with professionals to try and get as much relief as is available. A caring physician is such an important part of your team.

Helpful Tips to Remember

- Find a health care team that is sympathetic, helpful, and will continue to work with you through thick and thin.

- Understand that your health care team does not know all the answers and may be as frustrated as you are. Often, a team of health care professions is needed to address problems encountered by someone suffering from CPP. Psychosocial support is often necessary, as depression can also be present.
- Stay in touch with family and friends. Don't become isolated.
- Involve your family in treatment decisions.
- Remember that CPP is only one part of your life. Don't allow it to consume you.
- Talk to others about their experiences and ways of coping.
- Timed voiding (urinating by the clock rather than going to the bathroom whenever you feel the urge) is important. Over time, the bladder reeducates itself as the intervals between voids are lengthened.
- Wear loose clothing. Avoid belts or clothes that put pressure on your abdomen.
- Avoid activities that you know will cause a flare-up of your CPP: certain body positions, very bumpy rides, etc.
- Reduce stress. Try methods such as visualization, meditation, biofeedback, and low-impact exercises.
- Stop smoking. Nicotine, the active ingredient in tobacco, may worsen any painful condition, and smoking is harmful to the bladder.
- Try pelvic floor physiotherapy. Gently stretching and strengthening the pelvic floor muscles, with help from a pelvic floor physiotherapist, may reduce

muscle spasms. The physiotherapist may combine this technique with biofeedback.

- Treatment of the muscles and connective tissue in the pelvis by a highly trained physical therapist is highly beneficial for both CPP patients and those having other pelvic pain conditions. Instead of focusing on squeezing the muscles, such as during Kegel exercises, therapists teach the patient how to relax these key muscles. During therapy, the specialist focuses on the release of muscle tension and pain by manipulating the connective tissue and muscle in certain directions. Therapists often supplement their treatment with mild electrical, heat, and/or cold protocols.

The Rest of Vlad's Story

So, what happened to Vlad, who was held hostage by his chronic pelvic pain? Vlad had seen ten health care providers: three urologists, two primary care doctors, a naturopath, a chronic pain management specialist, a neurologist, a psychiatrist, and a dietician. He was suffering on a daily basis, describing his pain as "a seven to eight out of ten." He often had to take opiate medication for pain relief. He was experiencing depression and loss of productivity at his work.

Eventually, Vlad was able to find a compassionate physician who specializes in chronic pelvic pain; he was reassured that his problem was not a psychiatric one and that he wasn't crazy. Vlad was able to work with the pain management specialist to discontinue the use of opiate pain medication. He *was advised to stop bladder and prostate irritants such as caffeine, carbonated beverages, citrus fruits, and juices. He joined one of the support groups in his community, and he knew he wasn't alone and that there were others like him. Vlad had several bladder instillations of DMSO by his urologist and was able to use meditation and self-talk to relieve some of his urinary symptoms. Over a period of several months Vlad was able to gain control of his symptoms, and he described his pain as "two to three," which "made life worth living."*

The Bottom Line

Chronic pelvic pain in men is a common problem in many primary care practices, and even more common in urologic practices. Still, it remains underdiagnosed. Because the symptoms are non-specific and can be very confusing, men are sometimes thought to have psychological problems or are thought to have a bacterial infection, and they are treated with multiple courses of antibiotics despite the absence of real evidence of any bacterial infection.

The key to the correct diagnosis is awareness of the condition and its characteristics. The bottom line is that achieving a complete cure for chronic pelvic pain may not be possible; however, it is in reach of most men to achieve long-term remission, or certainly a reduction in symptoms that makes life more enjoyable.

CHAPTER 11

Urinary Incontinence

Ebenezer, a sixty-nine-year-old man, had prostate cancer which was surgically removed through a robotic procedure (see Chapter 3.) After surgery, he had a problem with urinary incontinence and had to wear diapers. The incontinence was a source of despair and even mild depression. It affected nearly every aspect of his life. He and his wife could not sleep in the same bed because of incontinence, and sexual intimacy was impossible.

From the time we are toddlers until we become older and infirm, we tend to take control of urination for granted. Seldom do we have embarrassing accidents between ages three and eighty-three. However, as men age, the muscles holding the urine in place until they can reach a restroom are weakened, and loss of urine becomes one of aging's most troublesome and devastating experiences. This chapter will discuss the problems that many middle-aged and older men have and what can be done to help.

Loss of urine is no laughing matter. It affects nearly fourteen million American adults. Let's put the myth aside that urinary incontinence (UI) only affects women. Incontinence is surprisingly common in men of all age groups. Losing urine is only half as frequent in men as in women; nevertheless, that represents a sizeable number of sufferers. Unfortunately, men with incontinence rarely discuss their problem with their physician, so the necessary attention is not paid to finding a cure to the condition or even to managing the problem.

Surprisingly, 25 percent of men aged forty or below reported incontinence at least once during the previous twelve months. At least 30 percent of men over age forty and 36 percent of men aged sixty to seventy had at least one incident of incontinence in the past year.

A Brief Review of the Male Plumbing System

The kidneys function to filter out liquid waste from the blood. The kidneys then combine this liquid waste with water and transport the urinary fluid to the bladder, where the urine is stored until it's time to pee. When the bladder muscle contracts, the external sphincter relaxes. Bladder contraction allows urine to exit through the urethra, the tube in the penis that transports the urine to the outside of the body.

Urinary incontinence occurs under one of three conditions: the bladder muscle does not contract; the bladder muscle contracts too frequently and without the man's permission; a weakened sphincter does not maintain the urine inside the bladder until it is time to empty it of urine.

Risk Factors for Incontinence in Men

Age: Men are more likely to develop UI as they grow older. This may be the result of physical changes that make holding

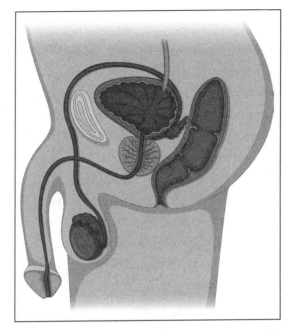

Figure 1: Male Pelvic Anatomy *(Shutterstock)*

urine more difficult. Certain diseases or conditions become more common with older age, and loss of bladder control may be an associated symptom.

Lack of physical activity: Being physically active may increase urine leakage, but not being physically active increases your risk for weight gain and decreases overall strength. This may make symptoms of UI worse.

Obesity: Extra weight on your midsection can place unnecessary pressure on your bladder.

History of certain conditions: Prostate cancer, an enlarged prostate, and treatments for these conditions can lead to temporary or permanent incontinence. Diabetes can also lead to UI.

Neurological issues: Diseases such as Parkinson's disease, Alzheimer's disease, and multiple sclerosis can interfere with your brain's ability to properly signal your bladder and urinary tract to store and expel urine.

There are many additional causes of urinary incontinence. These include urinary tract infections and prostate problems. Men who have had prostate surgery, bladder surgery, or are taking medications for urinary issues are two to three times more likely to experience incontinence. Table 1 provides a list of drugs that may cause urinary incontinence.

Drinking coffee or tea or taking prescribed medications with adverse side effects can aggravate your bladder. Certain drugs (e.g., diuretics, alcohol) have no direct action on the urinary tract but may contribute to incontinence by increasing urine production or impairing nervous system function.

Table 1 **Drugs that can cause urinary incontinence**
Alpha-adrenergic blockers
Antihistamines
Angiotensin-converting enzyme inhibitors
Calcium channel blockers
Diuretics
Analgesics/opioids (Percocet, Oxycontin)
Skeletal muscle relaxants
Antidepressants
Anticholinergics
Sedatives

Not surprisingly, daily or frequent incontinence is associated with deterioration in a man's quality of life. For instance, emotional health, social relationships, physical activity, and travel are less satisfactory for incontinent men. The problem also impacts a man's relationship with his significant other and his family and friends.

Male incontinence is a serious problem across all age groups in that it affects men's quality of life. Unfortunately, only a third of the men with incontinence discuss the problem with their physicians. However, three-quarters of them express an interest in evaluating and treating the problem once it is offered. Clearly, there is much room for improvement in diagnosis and management.

Types of Incontinence

There are five common types of urinary incontinence: stress incontinence, overflow incontinence, urge incontinence, mixed incontinence, and functional incontinence.

Stress Incontinence

Stress incontinence in men is usually a result of prostate surgery for either benign prostate enlargement or surgical removal of the prostate gland for cancer of the prostate. If you have stress incontinence, you leak small amounts of urine when you cough, sneeze, exercise, or put pressure on your bladder, and the muscle or sphincter cannot hold the urine in the body.

Urge Incontinence

Urge incontinence occurs when your bladder suddenly contracts and expels urine. You have an urge to urinate even though you know you emptied your bladder before. You urinate and then get the desire again a half-hour later. Urge incontinence often comes in waves. For example, it may not bother you all morning, but it becomes insistent mid-afternoon. You may feel the urge four or five times in the course of a few hours.

Mixed Incontinence

Mixed incontinence occurs when you have symptoms of two or more distinct types of incontinence, usually stress and urge incontinence. The symptoms of mixed incontinence are loss of urine with coughing and sneezing (stress component) and having to go to the restroom immediately with the first desire or urge to urinate (urge component). Your doctor needs to know the type of incontinence, as the treatments for each kind of incontinence may differ.

Overflow Incontinence

Overflow incontinence is a form of urinary incontinence characterized by the involuntary loss of urine from an overfull urinary bladder, often without any urge to urinate. Overflow incontinence occurs when you cannot completely empty your bladder; this leads to overflow, which leaks out unexpectedly. You may or may not sense that your bladder is full. The leakage, which can cause embarrassment and discomfort, is not the only problem. Urine left in the bladder is a breeding ground for bacteria. It can lead to repeated urinary tract infections and possibly kidney damage. The most common cause in men is an enlarged prostate, which impedes the flow of urine out of the bladder and results in constant dribbling of small amounts of urine both during the day and night.

Functional Incontinence

Functional incontinence is a form of urinary incontinence in which a man is usually aware of the need to urinate. Still, he cannot reach the restroom in time for one or more physical or mental reasons, and loss of urine occurs. Often, the cause of functional incontinence is a problem that keeps the man from moving quickly enough to get to the bathroom, pull down his pants to use the toilet, or transfer from a wheelchair to a toilet. Problems with mobility such as arthritis, back pain, or neurological problems, such as those which occur with Parkinson's disease or after a stroke, may also cause functional incontinence.

The Important Voiding Diary

A voiding diary helps the doctor achieve an accurate diagnosis and helps create a treatment plan. The doctor can better understand a man's incontinence problem if he has kept a voiding diary for just a few days before his examination by the doctor. Many men provide an unclear voiding history when simply relying on their memory; a voiding diary with real data is more accurate and helpful.

Date	Time	Urinated in Toilet	Small Accident	Large Accident	Fluid Intake	Circumstances of Accident

Figure 2: Voiding Diary

The simplest voiding diaries ask patients to record the frequency of incontinence episodes. Still, diaries also can be used to assess the situations in which incontinence occurs and to clarify the type of incontinence. For example, the diary may reveal leakage during times of increased abdominal pressure, suggestive of stress incontinence, or dribbling indicative of overflow incontinence. Men with stress incontinence usually wake once nightly or not at all. Patients with urge incontinence usually wake more than twice and as often as every hour to urinate.

The Physical Examination

The physical examination can identify clues that help lead to a diagnosis. The doctor will examine the abdomen and see if the bladder can be felt above the pubic bone, which is suggestive of overflow incontinence or a bladder that is not emptying completely. The lower legs are examined for joint impairment, which might indicate incontinence secondary to an inability to reach the toilet quickly (functional incontinence). Swelling of the lower extremities may indicate heart disease and that incontinence results from the excessive fluid that overwhelms the kidneys and bladder. Any physical must include a prostate examination or digital rectal examination.

In addition to the history, the voiding diary, and the physical exam, minimal testing may be necessary. The additional testing includes urinalysis, a urine culture, a cystoscopy, and perhaps urodynamic testing. The latter test includes measuring the bladder capacity, the flow of urine from the bladder to the outside of the body, and any urine retained in the bladder. A few blood tests check kidney function, calcium, and glucose levels.

Treatment of Urinary Incontinence

Diapers or pads are often the first-line treatment for mild to severe incontinence. There are ultra-absorbent and comfortable products that look and fit like regular underwear. These products are specially constructed to prevent leaks and quickly move liquid away from the body while preventing any odor. The soft, cloth-like material is comfortable against the skin, while added spandex creates a sleeker, more form-fitting design that can't be seen under clothes. A growing number of pads for day and night use and absorbent underwear and bed pads are available at pharmacy chains, on the internet, and at discount stores. Wearing a diaper or pull-ups may be the difference between being confined to your home and feeling able to leave the house and participate in social activities. Nevertheless, most men who use diapers are interested in finding other solutions to the problem of incontinence.

Every man can make simple lifestyle changes that may diminish his urinary incontinence.

Low impact sports such as cycling, yoga, or elliptical machine exercises are ideal activities for keeping fit without affecting a sensitive bladder condition.

Abdominal workouts such as sit-ups, crunches, or plank kicks place a lot of pressure on the pelvic floor. Use alternative exercises where breathing or the position supports the pelvic floor. Finally, avoid heavy lifting. Lifting heavy objects is particularly bad for the pelvic floor and back. Ask for help instead.

Pelvic Floor Exercises

Pelvic floor exercises and targeted Pilates and yoga exercises can strengthen the pelvic muscles. Practicing at least three times a day can help support the pelvic floor muscles and give more control when needed.

Kegel exercises may help strengthen the muscles in your pelvic floor. This allows you to delay urinating until you reach a toilet. You may have thought that Kegel exercises were something only women do. In fact, the muscles strengthened with Kegel exercises are the same in both sexes.

Men often have a weakness in the pelvic floor muscles after surgery that requires treatment and strengthening. Therapists will utilize bladder training, biofeedback, pelvic floor muscle exercises, and electrical muscle stimulation—which is not painful but encourages contraction and relaxation of the muscles responsible for holding the urine in the bladder. Biofeedback is often prescribed for men with stress incontinence following prostate gland surgery for prostate cancer.

Men typically attend in-office appointments two to three times per week for twelve weeks. Treatment requires a digital rectal exam evaluating the external and internal tissues of the pelvis, especially those surrounding the urethra. This type of therapy is effective in some men, especially if they complete the course of treatment and do the exercises between treatment sessions.

Drink Just Enough

There's no need to avoid consuming fluids to reduce the urge to visit the bathroom. Limiting water intake makes urine more concentrated, which boosts the chances of bladder irritation. Just say no to caffeine. Caffeine, alcohol, and carbonated drinks could be your new worst enemies. Try limiting coffee, tea, and carbonated beverages for a week or two as they can irritate a sensitive bladder.

Your bladder is trainable. Set your alarm on your smartphone to alert you to use the restroom. If you need to pass water frequently and rush to the bathroom, ask your doctor about a daily schedule for building up the bladder's holding capacity. Remember, allow your bladder to empty completely each time you go to the toilet.

Condom Catheter

A condom catheter is a treatment option for men who have a constant loss of urine or cannot receive any warning that urination is imminent. A condom catheter consists of a condom with the tip of the condom cut off, so it can be attached to a drainage tube and collecting bag, which is often worn on the leg or thigh.

Condom catheters are made in three sizes: small, medium, and large, and they usually contain adhesives or skin sealants, which are non-irritating to the underlying skin. The skin sealant protects your skin from sweat and urine irritation; when you remove the condom catheter, the layer of skin sealant is removed, not the external layer of your skin.

Penile Clamp

A penile clamp uses a hinged, rigid frame that supports two pads and a locking mechanism. It controls leakage by applying constant pressure on the penis and compresses the urethra, the tube in the penis that transports urine from the bladder to the toilet. The penile clamp cannot be used for long periods and should not be placed on the same location of the penis since the clamp decreases the blood supply to the underlying skin and urethra and can result in damage to the urethra. We recommend the penile clamp be moved up or down the shaft of the penis every two hours and not used while sleeping. For the most part, the penile clamp is a temporary solution, and a man with incontinence should not become dependent on this treatment option.

Medical Management

Medication is possible for men with incontinence related to prostate gland enlargement or urge incontinence. The medications finasteride and dutasteride reduce the size of the prostate. Incontinence secondary to an enlarged prostate can be managed with medications such as alpha-blockers that relax the prostate muscles. Other options for the enlarged prostate gland that doesn't respond to medications are described in Chapter 2.

Medications to address urge incontinence relax the bladder, reducing the urges and the subsequent incontinence. These anticholinergic medicines, such as oxybutynin, control bladder muscles and increase bladder capacity. Antidepressant medicines, such as imipramine, may also help with bladder control.

There are very few medications that are effective for stress incontinence. If a man has both stress and urge incontinence, then anticholinergics may be helpful with the urge component of the incontinence. Also, the antidepressant medication, imipramine, may be beneficial with mild stress incontinence. Botulinum toxin type A (Botox) can be injected into the bladder to relax the bladder muscles.

Surgical Options

Male Slings

A small sling made of soft synthetic mesh can be implanted inside the body to support surrounding muscles. This can help keep the urethra closed, especially when coughing, sneezing, and lifting.

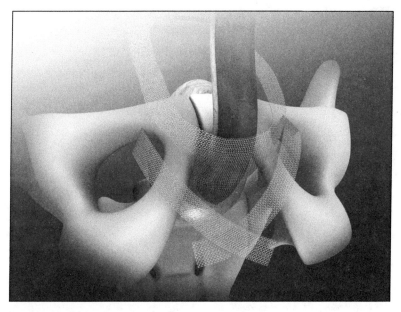

Figure 3: Male Sling *(Coloplast)*

Bulking agents

A synthetic material is injected around the urethra. This material will compress the urethra and help the urethra close when you are not urinating.

Artificial Sphincter

The artificial urinary sphincter (AUS) is placed inside the body. The AUS is designed for use after prostate surgery. A saline-filled cuff compresses the urethra and prevents urine from leaving the bladder. The AUS also contains a pump in the scrotum that can be activated to remove fluid from the cuff, releasing compression on the urethra to allow urination when the man feels his bladder is full.

Advice Regarding Sexual Activity

Some men with urinary incontinence avoid sexual intimacy. You can still have sex, but we suggest a few precautions prior to engaging in sexual intimacy:

1. Avoid caffeine and alcohol for several hours prior to sex.
2. Avoid all liquids one hour before sex.
3. Empty your bladder as completely as possible before sex.
4. Place a towel between you and your partner and the bed if you are concerned about leaking during sex.

What Ever Happened to Ebenezer?

Ebenezer used diapers and an external condom catheter for several months, practicing Kegel exercises diligently every day. He gained considerable control over his incontinence. He avoided urinary accidents by making an effort to urinate before leaving his home. He used a tiny pad for any dribbling when his bladder was full. Ebenezer returned to his bedroom with his wife, and his depression lifted. He took oral medication for his erectile dysfunction, and his PSA tests after surgery were 0.01ng\ml, which means he was cured of his prostate cancer. Ebenezer got his life back, and both he and his wife are happy campers.

Bottom Line

Most male issues with urinary incontinence can be improved by several routes: adopting simple lifestyle changes, practicing pelvic floor exercises, and/or finding the right products that promote a more normal social life. Many can also be cured through a range of exercises, medication, and surgery. Sufferers need not suffer in silence; they must speak to their physicians and health providers, insisting that they pay closer attention to their urinary concerns.

CHAPTER 12

Sexually Transmitted Infections

When Charlie, age twenty, woke up in his dorm room, he had a very strong urge to urinate—more than most mornings. Expecting relief as soon as he reached the bathroom, he was surprised by the painful burning as he emptied his bladder. When he looked down, he noticed a yellowish discharge in the front of his underwear. Since it was the weekend, the campus health center was closed. He did not know where to turn. He certainly could not call his parents. Although he was sexually active, he did not have a serious girlfriend in whom he could confide. His best thought was to turn to the internet for a solution.

Introduction

When most people think of sexually transmitted diseases, they think of herpes and HIV. As if those are not enough to worry about, there are a host of viruses, bacteria, and other organisms that can be transmitted sexually. Not only can these infections be spread through vaginal and anal contact, but also through oral and manual (hand) contact. In some cases, these infections can cause some very bothersome symptoms. At other times, symptoms can be delayed. Often, untreated sexually transmitted infections (STIs), also called sexually transmitted diseases (STDs) can cause irreversible damage to the genitourinary and other organ systems. The good news is that most of these diseases are preventable. It all starts with knowledge.

Prevention by Intention

Other than abstinence, the best way to prevent acquiring a sexually transmitted infection is to commit to a monogamous relationship with an uninfected partner as determined by their physician. When a monogamous relationship is not possible, sex with as few partners as possible is the next best thing.

Of course, the most common method of prevention when the status of infection is unknown is condom use. However, condoms are only effective when used properly. First and foremost, any penile contact should involve a condom. This hard-and-fast rule not only applies to vaginal and anal contact, but also to oral and manual contact. Both semen and vaginal secretions of infected individuals can be highly contagious. Even "casual" contact—either before or after sexual intimacy—can lead to unsuspected exposure to these secretions.

Although somewhat controversial in the public's eye, the Centers for Disease Control and Prevention has well-established recommendations for vaccination.[1] Every man should be vaccinated for HPV and Hepatitis B early in life. These vaccinations will be discussed later in this chapter.

Early Detection

Seek medical attention at the first sign of symptoms. These symptoms can include genital or pelvic pain, skin lesions, urethral discharge, vaginal discharge, pain with intercourse (especially for females), and pain with urination. Any of these concerns should be shared with your sexual partner or partners so that they can be properly treated. In some circumstances, diagnosis can be easier in your partner, thereby avoiding a missed diagnosis.

With the exception of a mutually monogamous relationship in which both partners are confirmed to be uninfected, regular testing is the hallmark of prevention and early detection. Most people do not realize that many sexually transmitted infections, such as HIV and syphilis, may not cause any symptoms, particularly at the onset. More importantly, some of these infections can cause irreversible damage to the genitourinary tract and other organ systems. Appropriate testing can help prevent some of these long-term consequences.

Late Complications of Sexually Transmitted Infections

Several of the sexually transmitted viruses are now known to cause a variety of cancers. Human papillomavirus (HPV) can cause cervical, mouth, throat, and anal cancer. Hepatitis B and C can cause liver cancer. The Centers for Disease Control and Prevention recommends vaccination for both HPV and Hepatitis B. Untreated human immunodeficiency virus (HIV/AIDS) can lead to an assortment of cancers and death. Early treatment of HIV can help prevent cancers associated with this disease.[2]

Since dealing with an STI is a couple's issue, a man should also be concerned about the well-being of his partner—especially when it can affect future fertility. Sexually transmitted infections have a much higher risk of future fertility problems in women than in men.

These infections can cause pelvic inflammatory disease, also known as "PID." Chlamydia and gonorrhea are the most common causes of PID and can lead to scarring and blockage of the fallopian tubes (the tubes that carry the eggs to the uterus). Similarly, a man's tubes can also develop scar tissue from untreated gonorrhea and put him at risk for future fertility problems. Furthermore, any active sexually transmitted infection during pregnancy can threaten the well-being and life of the fetus.

The Hidden Culprits

All sexually transmitted infections can be treated, and most can be cured. Knowledge is power when it comes to fighting these bacteria, viruses, and other organisms. Unfortunately, you may not notice any symptoms with many of these infections. For this reason, you should see a physician for testing when starting a sexual relationship with a new partner or if there are any concerns of having a sexually transmitted disease. What follows is a guide to the most common sexually transmitted infections, including typical symptoms, methods of diagnosis, and the standard treatments.

Genital Warts

Human papilloma virus, or HPV, is the causative organism of genital warts. These warts can appear anywhere along the penis, scrotum, anus, inner thigh, vagina, and cervix. This virus has dozens of subtypes, some of which do not cause the classic wart associated with this disease. The virus can be transmitted with oral, genital, anal, and even skin-to-skin contact. Other than through a visual inspection, the diagnosis can be made with biopsy of the skin lesion or swabbing of the cervix.

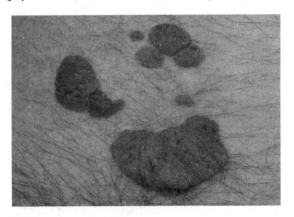

Figure 1: Genital Warts

Not all types of HPV are harmful, but some can lead to penile, anal, mouth, and throat cancer. Prevention, early detection, and proper treatment are paramount in avoiding these cancers. The CDC recommends that young men ages eleven to twenty-one be vaccinated for HPV.[3]

Many of the subtypes of HPV are not symptomatic and are eradicated by the body's immune system. Visible skin lesions usually appear within two to three months following infection and are most commonly treated with chemical or laser destruction, depending on the extent and location. In some cases, your physician may prescribe one of two available topical creams, Condylox® (podofilox) or Aldara® (imiquimod). These creams can take many weeks and can cause irritation of the surrounding skin. Although the creams clear the warts in roughly 50 percent of patients, 20-50 percent will recur. Laser destruction is usually performed in an operating room setting.

Gonorrhea and Chlamydia

Although caused by different organisms, gonorrhea and chlamydia often coexist in an infected individual. In fact, even when only one of these diseases is detected, treatment is given for both. The most common symptoms are burning in the urethra (worse with urination) and a milky or yellowish discharge from the urethra. Like most sexually transmitted infections, these organisms can be transmitted through vaginal, anal, and oral sex. More than 350,000 cases of gonorrhea and more than 1.5 million cases of chlamydia are reported each year.[4] Urethral swabbing and testing of any urethral discharge usually confirms the diagnosis. It is very important that your partner also be treated and that you refrain from sexual contact until both you and your partner have both been fully treated.

Most men with gonorrhea have symptoms consisting of burning on urination, frequency of urination, and a urethral discharge. Taking into account the growing resistance of these organisms to certain antibiotics, the CDC recommends dual therapy with a single intramuscular dose of ceftriaxone (250 mg) plus a single oral dose of azithromycin (1 gm). Once again, both organisms, gonorrhea and chlamydia, must be treated, even when only one is detected by testing.[5]

The "Gift" That Keeps on Giving

Herpes viruses come in many forms and can cause a variety of diseases including chicken pox, shingles, cold sores, mononucleosis ("mono"), and genital herpes. Unlike HPV, herpes is not usually associated with a future risk of cancer, although the subtype that causes mono *is* associated with lymphoma.

The herpes viruses that are transmitted sexually are called herpes simplex and are classified in two types. Herpes virus type 1 (HSV-1) is usually associated with oral blisters or "cold sores," whereas herpes virus type 2 (HSV-2) is usually associated with genital lesions. However, either type can occur in or around the mouth, penis, anus, and rectum. The lesions are characterized by painful blisters.

Although these lesions are treatable with antiviral medications, there is no cure to prevent recurrences. For those with recurrent disease, a characteristic pain or sensation can precede blister formation by several days. During both this "prodrome" phase and when the lesions are visible is the time during which you are most contagious. Herpes is very contagious and can be transferred by any affected skin not covered by a condom. Although there is no surefire way to prevent transmission to an uninfected partner, any person with HSV should consult their physician for advice on how to minimize this risk.

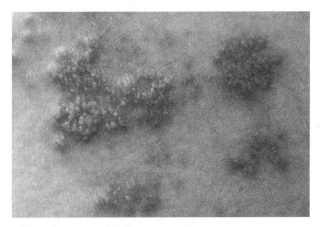

Figure 2: Herpes Skin Lesion

The diagnosis of HSV can usually be made by visual inspection of the lesion by your physician. At times, your physician may swab the sore for laboratory testing. You should also observe any changes in the appearance of the blisters and any sensations or discomfort that preceded the lesion. Often, the blister starts with a small red patch that contains multiple pin-head-size blisters that soon coalesce to form one blister. These are usually painful and tender to the touch, but some cause no pain (especially inside the vagina or rectum). The blisters will then break open and form an ulcer before forming a scab during the healing phase.

As mentioned, herpes simplex is treatable but not curable. In fact, once contracted, it lies dormant in your nervous system for your lifetime. In some cases, years pass between exposure and your first outbreak. Subsequent outbreaks as well can be separated by years, though more frequent outbreaks are not unusual. Treatment focuses on quicker healing of sores, lessening the severity and frequency of outbreaks, and minimizing transmission of the virus to an uninfected partner (again, not fool-proof).

Three oral medications are currently available—acyclovir, valacyclovir, and famciclovir. Although topical antiviral creams have been used in the past, they have since been shown to be ineffective. The CDC recommends a seven to ten-day course of oral antiviral medications in all patients with their initial outbreak. Shorter courses can be used for subsequent outbreaks. In fact, for those who can identify the prodrome (early signs or symptoms) of abnormal sensations that precede blister formation, a one- to three-day regimen can be very effective if started promptly. In these cases, a supply of pills should be on hand for effective "self-start" therapy.[6]

Another approach is suppressive or preventative (prophylactic) therapy. Daily low-dose administration (reduced by as much as half) may be appropriate for those with frequent outbreaks. The frequency of outbreaks often decreases with time, and therefore the need for daily dosing should be re-evaluated on an annual basis with your doctor. However, the risk of transmission to an uninfected partner can be dramatically reduced with continuous daily dosing, particularly in conjunction with condom use. These benefits should be carefully weighed against the disadvantages of long-term use such as cost, side effects, and future drug resistance.

Human Immunodeficiency Virus (HIV)

Over the past few decades, we have come a long way in the management of HIV (human immunodeficiency virus), the virus that causes AIDS. However, we have still not found a cure or an effective vaccine. This virus infects cells in our bloodstream that are responsible for a major part of the function of our immune system. As a result, those individuals infected with HIV are more susceptible to other infections and some cancers, many of which are otherwise rare. Although many people have managed their HIV infection for ten or more years with medications, this disease should be considered potentially fatal.

Infection with HIV often occurs alongside other sexually transmitted infections. In fact, anyone infected with HIV should be tested for all of the other diseases mentioned in this chapter (among others). In addition, the presence of another STI will make you more susceptible to HIV infection since STIs break down our natural barriers.

Regular testing should be done for those with multiple sexual partners, a new partner, and other high-risk exposure sources: needles, tattoos, health care providers). Considering the gravity of such an infection, trust in a relationship should not be underestimated.

Although there is no cure for this virus, tremendous advances in treatment have been made over the last few decades. The CDC currently recommends a "cocktail" of three different antiretroviral drugs from at least two different drug classes. New medications and combinations are constantly being evaluated, and for the individual with HIV it is important for them to be in fairly frequent contact with their treating physician. Many infected individuals can go many years with undetectable viral counts. Despite these low counts, care must be taken to avoid exposure to others.[7]

Syphilis: "The Great Imitator"

No discussion of sexually transmitted infections is complete without the inclusion of syphilis. It was called "the great imitator" because it was often confused with other conditions. Often, not much attention is given to syphilis because it is not as common as other STIs and because it is curable. However, we are starting to see a rise in the incidence of this disease.[8] In fact, the CDC recommends that all pregnant woman be tested since syphilis can be easily passed from mother to baby.[9] More importantly, if not treated in the early stages, syphilis can lead to some serious health and life-threatening conditions.

Syphilis is divided into four stages—primary, secondary, latent, and tertiary. Primary syphilis is characterized by small, round, and painless bumps (called "chancres") in the area where the organism (the "spirochete" bacterium *Treponema pallidum*) originally entered the skin. Transmission usually occurs via the penis, vagina, anus, rectum, or mouth. Whether or not treatment is received at this time, the chancres last three to six weeks.

During or shortly after the primary stage is healing, secondary syphilis will manifest with a flat, itchy rash that can occur on any part of the body (most commonly on the palms or on the soles of the feet). Often, this rash can have minimal symptoms and can be difficult to diagnosis unless syphilis is being considered. Like primary syphilis, secondary syphilis will resolve whether or not it has been treated.

The latent stage of syphilis can last for years without any signs or symptoms. However, if it is never properly treated, it can unpredictably lead to the dangerous tertiary stage. Tertiary syphilis can have life-threatening effects on the brain (neurosyphilis), eyes (ocular syphilis), and other internal organs.

As already mentioned, the key to diagnosis is having the suspicion of the possibility of syphilis. Often the patient's description of events and risk factors can be helpful, and at times the spirochete can be sampled with a swab to confirm its presence with laboratory testing. However, the diagnosis is usually made with a special blood test called an RPR. Syphilis is simply treated with penicillin. For those allergic to penicillin, doxycycline is usually effective. Of course, all sexual partners from the time of initial infection must also be treated.

The Crabs

There is nothing like an unrelenting jock-itch to make you "crabby." However, unlike the traditional jock-itch that is caused by a fungus (as is athlete's foot), crabs is caused by lice, a parasitic insect. It is very contagious—not only through sexual contact, but also through non-sexual contact with clothes and other inanimate objects, such as pillows.

Since the pubic hair is exposed, condoms are usually not effective (even with "manscape" shaving). Three million cases are reported in the US each year.[10]

Diagnosis can usual be made by identifying the small insects with the naked eye or easily with a magnifying glass. Treatment is simply a one-time application of an over-the-counter lotion, 1% permethrin, to the entire body with special attention to areas with hair (including the eyebrows). Your pharmacist can help guide you on selection and use of this product. Also, it is best to reapply the lotion seven days later to treat any lice that may have hatched following the treatment, but prior to them being mature enough to reproduce. Even though the lice do not survive longer than twenty-four hours away from their human host, it is important to thoroughly clean any clothing, sheets, towels, or surfaces that may contain lice in order to prevent reinfection shortly after treatment (or infection of others.)

Trichomoniasis

Trichomoniasis is caused by a parasite and can manifest in itching, burning, malodorous urethral discharge, or no symptoms at all. Transmission occurs most commonly via direct genital contact but can also occur via secondary contact with objects such as sexual paraphernalia. Treatment is simply a single oral dose of metronidazole 2 gm (Flagyl) for both partners.

Hepatitis

When talking about sexually transmitted infections, most people do not think about liver disease or hepatitis. However, in addition to transmission through blood (needles and transfusions), the various types of hepatitis can be passed via food (especially Hepatitis type A), saliva, and other body fluids. In fact, with most types, sexual contact is one of the more common causes of transmission. Since some of these viruses often coexist with HIV, a person diagnosed with one of these viruses should be tested for the other.

Common symptoms include fatigue, nausea, fever, and yellowing of the skin and the whites of the eyes. Hepatitis type A (HAV) is usually transmitted through poorly kept or poorly processed food (seafood being the most common), but it can also be transmitted sexually. Type A usually goes away on its own, but for those at higher risk, a vaccination is available.

The vaccination is often given to global travelers who go to areas that have a high incidence of Hepatitis A. The CDC recommends incorporation of HAV vaccination into the childhood vaccination schedule.

Hepatitis B (HBV) is transmitted through blood transfusions, contaminated needles, and sexual contact. It is usually treated by careful observation of liver function, monitoring of blood viral levels, and other supportive care. The CDC recommends vaccination of all children, health care workers, and those at higher risk for infection. The CDC hopes that the entire population would be vaccinated for both HAV and HBV so that these menacing viruses can be eradicated from our communities. Although there is no cure for HBV, there has been some promising results with treating chronically infected patients with immunotherapy to control the disease.

Hepatitis C (HCV) has the same risk factors, transmission methods, and potential consequences as HBV. However, there is a cure for HCV that is FDA-approved and commercially available (direct-acting antivirals.)

Hepatitis D (also called Hepatitis Delta or HDV) interestingly must coexist with Hepatitis B in order to survive.

Unfortunately, the combination of these two viruses gives a much higher rate of permanent liver damage (70 to 80 percent) and death (20 percent.) Therefore, vaccination for HBV prior to infection with either of these viruses is the most effective way to prevent HDV.

For any of the types of hepatitis, prevention is best achieved with proper vaccination, avoidance of contaminated needles, and sensible sexual practices. If you are exposed to a person or substance at risk for hepatitis, seek medical attention immediately. In some cases, postexposure prophylaxis can stop an infection in its tracks before it has a chance to establish itself in your body.[11]

Molluscum Contagiosum Virus

Molluscum contagiosum is a viral infection that causes benign skin lesions called papules. The virus can be transmitted by sexual contact as well as by casual contact and can be seen in children and adults. It is an uncommon but well recognized condition. The lesions in the genital area are usually multiple and elevated, and they can measure 2-5 mm in size. In most cases there will be multiple lesions. Molluscum can be asymptomatic or can cause itching and skin irritation. These lesions can sometimes resolve without treatment, but there are topical medications that can be used. Dermatologists will often use a scraping technique called curettage to eradicate the lesions. Though not a serious condition, it is transmittable by sexual contact and partners should be examined.[12]

So, Whatever Happened to Charlie?

Unfortunately, Charlie's search for an answer on the internet did not yield a home remedy. However, he did find an urgent care center nearby that was open on the weekend. After hearing Charlie's symptoms, the physician suspected that Charlie suffered from urethritis due to a sexually transmitted infection. The doctor placed a small swab into Charlie's urethra in order to take a sample of the secretions for laboratory analysis. The initial examination under the microscope revealed a unique type of white blood cells, polymorphonuclear leukocytes. Since these are characteristic of gonorrhea, he was treated with "dual therapy" of a ceftriaxone injection and a single oral dose of azithrothromycin. A few days later, the urgent care center called Charlie to let him know that the culture results revealed gonorrhea and chlamydia, both of which should be covered by the antibiotics that he received. In fact, Charlie's symptoms were almost gone. During the last couple of months, he only shared sexual intimacy with one partner. He kindly shared this information with her so that she could be properly examined and treated.

The Bottom Line

Prevention of sexually transmitted infections begins with an open and honest conversation between two partners. Prior to any sexual activity (vaginal, oral, or anal), both partners should undergo testing. Ideally, both partners should commit to having sex only with one another.

In situations outside of a committed relationship, you should be sexually active with as few partners as possible and always use a condom. Prompt attention to any symptoms, testing between sexual relationships, and appropriate vaccination are the hallmarks to prevention in these circumstances.

References

1. Centers for Disease Control and Prevention. "Child and Adolescent Immunization Schedule by Age." Last reviewed April 27, 2023. Accessed August 1, 2017. https://www.cdc.gov/vaccines/schedules/hcp/child-adolescent.html.

2. American Sexual Health Association. "STIs and Cancer." Accessed August 1, 2017. http://www.ashasexualhealth. org/stis-and-cancer.

3. Centers for Disease Control and Prevention. *Sexually Transmitted Disease Surveillance 2015*. Atlanta: U.S. Department of Health and Human Services, 2016. https://www.cdc.gov/std/stats/archive/STD-Surveillance-2015-print.pdf.

4. Centers for Disease Control and Prevention. *2015 Sexually Transmitted Diseases Treatment Guidelines*. Atlanta: U.S. Department of Health and Human Services, 2015.

5. Centers for Disease Control and Prevention. "Gonococcal Infections among Adolescents and Adults." Last reviewed September 21, 2022. https://www.cdc.gov/std/treatment-guidelines/gonorrhea-adults.htm.

6. Centers for Disease Control and Prevention. "Genital Herpes" Centers for Disease Control and Prevention. Last reviewed September 21, 2022. https://www.cdc.gov/std/treatment-guidelines/herpes.htm.

7. National Institutes of Health. "HIV Clinical Guidelines: Adult and Adolescent ARV—What's New in the Guidelines?" March 23, 2023. https://clinicalinfo.hiv-test.od.nih.gov/en/guidelines/hiv-clinical-guidelines-adult-and-adolescent-arv/whats-new.

8. Centers for Disease Control and Prevention. "Syphilis Statistics." Last modified April 11, 2023. Accessed August 1, 2017. https://www.cdc.gov/std/syphilis/stats.htm.

9. Centers for Disease Control and Prevention. "Syphilis During Pregnancy." Last reviewed July 22, 2021. Accessed August 1, 2017. https://www.cdc.gov/std/treatment-guidelines/syphilis-pregnancy.htm.

10. Chosidow, Olivier. "Scabies and pediculosis." *The Lancet* 355, no. 9206 (March 2000): 819–26 https://doi.org/10.1016/S0140-6736(99)09458-1.

11. Centers for Disease Control and Prevention. "Viral Hepatitis." Last reviewed March 8, 2023. Accessed September 1, 2017. https://www.cdc.gov/hepatitis/index.htm.

12. Chen, Xiaoying, Alex V. Anstey, and Joachim J. Bugert. "Molluscum Contagiosum Virus Infection." *The Lancet Infectios Diseases* 13, no. 10 (October 2013): 877–88. https://doi.org/10.1016/S1473-3099(13)70109-9.

CHAPTER 13

Gynecomastia and Male Breast Cancer

Sebastian, age forty, noticed progressive enlargement of both of his breasts. His breasts were not tender and there was no mass noted in either breast nor was there discharge from his nipples. This was a condition of concern to Sebastian, especially when he took off his shirt at the pool or the beach. What could he do?

Breast problems are often thought of a problem that only affects women, but in rare cases men can have breast problems. This chapter will discuss the two most common causes of breast enlargement: breast cancer and gynecomastia.

Breast cancer is one of the most common cancers in women. Women have mobilized and raised hundreds of millions of dollars and achieved publicity and public awareness for breast cancer. Every October the National Football League adopts a pink color for shoes and emblems on the uniforms of the football players to highlight breast cancer in women. However, breast cancer awareness is not promoted to men nearly as frequently, yet men are not immune to this problem. Women have been educated to perform self-exams on a regular basis and obtain mammograms every year after age forty-five. Many men, including some doctors, do not realize that men have breast tissue and that they can develop breast cancer. Also, men aren't instructed to perform self-examination of their breasts, whereas women are advised to conduct self-exams on a regular basis.

Breast cancer is about 100 times less common among men than among women. However, about 2,360 men develop breast cancer each year, and about 430 die from this cancer. Because male breast cancer is uncommon, it may not be suspected as a cause of symptoms. As a result, male breast cancer often progresses to an advanced stage before it is diagnosed.

Possible signs of breast cancer to watch for in men include a lump or swelling, which is usually (but not always) painless. Another sign is skin dimpling or puckering which may result in nipple retraction (turning inward). Finally, redness of the nipple area or a discharge from the nipple warrants further evaluation.

These changes aren't always caused by cancer. For example, most breast lumps in men are due to gynecomastia (a harmless enlargement of breast tissue). Still, if you notice any breast changes, you should see your health care professional for further evaluation. The diagnostic techniques for men are the same as those used in women: mammography, ultrasound, CT scan and, if indicated, a breast biopsy.

No discussion on breast cancer either in men or women would be complete without mentioning the BRCA gene. Men and women have BRCA1 and BRCA2 genes. The function of the BRCA genes is to repair cell damage and keep breast cells and other cells growing normally. But when these genes contain abnormalities or mutations that are passed from generation to generation, the genes don't function normally, and the risk of breast and prostate cancer increases. Men who have an abnormal BRCA2 gene have a higher risk of breast cancer—about eight times greater than average. Genetic testing for the BRCA gene consists of a blood or saliva test. As in women, treatment options for breast cancer in men include surgery, radiation therapy, and chemotherapy.

Most of the information about treating male breast cancer comes from doctors' experience with treating female breast cancer. Because so few men have breast cancer, it is difficult for doctors to study the treatment of male breast cancer patients separately in clinical trials.

Local therapy is intended to treat a tumor without affecting the rest of the body. Surgery and radiation therapy are examples of local therapies. Systemic therapy consists of drugs, which can be given by mouth or directly into the bloodstream to reach cancer cells. Chemotherapy, hormone therapy, and targeted therapy are also considered for advanced breast cancer, that is cancer that has spread beyond the breast to lymph nodes, bones, or other organs.

The prognosis for men with breast cancer was once thought to be worse than for women. Recent studies have not found this to be true. In fact, men, and women with the same stage of breast cancer have a similar outlook for survival.

Gynecomastia: Benign Breast Enlargement

Breast enlargement in males is called either gynecomastia or pseudogynecomastia. Gynecomastia is a benign condition consisting of enlargement of the breast tissue itself, which consists of glands. Pseudogynecomastia is the appearance of enlarged breasts in overweight men. However, this enlargement is usually the result of an increase in fat tissue around the

Gynecomastia can occur at these times:

- **Birth:** More than half of male newborns have enlarged breasts, or breast buds. This condition is due to the mother's estrogen levels. The enlarged breasts usually go away within a few weeks.

- **Puberty:** More than half of teenage boys have some degree of breast enlargement. Fluctuating hormones, including drops in testosterone and surges in estrogen, cause breast tissue to grow. The condition goes away as hormone levels even out — a process that takes about six months to two years to complete.

- **Adulthood:** Enlarged breasts are more common in men over fifty. With age, men's bodies produce less testosterone. They may also have more body fat, which stimulates estrogen production and breast tissue growth.

The enlargement is usually caused by these factors:
- liver or thyroid disorders
- obesity
- heavy use of marijuana, beer, alcohol, or heroin

Various medications may be responsible for gynecomastia:
- Antidepressants
- Bacterial and fungal infections

- Prostate for treating benign enlargement of the prostate gland
- Stomach pain treated with proton pump inhibitors
- Anti-hypertensive medications
- Male infertility treated with human chorionic gonadotropin, or HCG
- Substance abuse protocol (by taking methadone)
- Amphetamines, cannabis, and opioids
- Testosterone or other anabolic steroids used to build muscle
- Lavender oils and tea tree oils (found in herbal supplements and skin care products)

Generally, no specific treatment is needed for gynecomastia. Breast enlargement often disappears on its own or after its cause is identified and treated. Surgical removal of excess breast tissue is effective but rarely necessary. Liposuction, a surgical technique that removes tissue through a suction tube inserted through a small incision, is the preferred surgical option. Additional cosmetic surgery sometimes follows the liposuction.

Nipple Discomfort: Jogger's Nipple

Irritation of the nipple is more common than enlargement of the breast itself. Pain, redness, and even bleeding of the male nipple are common complications of intense, prolonged exercise—hence the common names "jogger's" and "marathoner's" nipple. The cause is not running itself but the mechanical irritation of the runner's shirt rubbing up and down against his chest, when running for long distances, especially in hot, humid weather.

You don't have to give up running to cure jogger's nipple. Instead, apply some petroleum jelly to your nipples *before* running. Plastic adhesive strips are also helpful in preventing jogger's nipple discomfort. One other option is to simply run without a shirt when it's hot or humid.

Sebastian's Solution

Sebastian made an appointment with his primary care physician who suggested a mammogram. Happily, the image did not suggest breast cancer. Sebastian was referred to an endocrinologist. The endocrinologist concluded that Sebastian had benign gynecomastia. Since Sebastian was moderately obese (BMI 30), the endocrinologist suggested a weight loss program including improved nutrition and an increase in exercise. In a few months after the weight loss regimen, the gynecomastia disappeared, and Sebastian was comfortable removing his shirt at the swimming pool.

Bottom Line

Breast conditions, including breast cancer, occur in men as well as women. While not as common in men as in women, men need to know that any breast lumps, swelling or discharge from the nipple should be examined by a physician.

CHAPTER 14

Diet and Nutritional Supplements

Sam is a forty-seven-year-old truck driver who is 5'8" and weighs 270 pounds. He suffers from high blood pressure that is controlled with medication. Sam is on the road most of the day and admits that he doesn't have time to think about his diet. He eats hamburgers, French fries, and pizza frequently. He also enjoys a few beers nightly once he gets to his destination. Sam's wife told him that he should take caffeine powder as a supplement for weight loss, but he wants some advice on proper nutrition.

You Are What You Eat

One of the most common questions that health care professionals receive is "What should I eat to stay healthy, prevent diseases, and build muscle?" Another common question is "What dietary supplements should I take?" This chapter will examine the foods and supplements that are beneficial for men's health and the supplements that should be avoided.

A Balanced Diet

If you are like most people, you have tried various diets to stay healthy, lose weight, or build lean tissue. Fad diets (like the cabbage soup diet, low-fat diets, low-carbohydrate diets, and others) may help with weight loss temporarily because they tend to restrict calories. However, as soon as we resume "normal" eating, the weight may come back, or even worse, we may gain more weight. This is what is called "yo-yo" dieting. For men (and women) to be healthy, we should eat a balance of carbohydrates, protein, and fat. Let's take a closer look at these macronutrients.

Carbohydrates

Carbohydrates are the body's main source of energy. More than half of our calories should come from "healthful" carbohydrates. However, not all carbohydrates are healthful. For example, sugar is a simple or refined carbohydrate. These types of carbs cause the body to increase insulin levels in the blood, which has been linked to weight gain. Simple or refined carbohydrates contain only one or two sugar molecules and are broken down and released into the bloodstream quickly. Other refined carbohydrates include white flour, potatoes, and white rice.

A healthy diet includes "complex" carbohydrates, which are made up of three or more sugar molecules. These take longer for the body to break down, so the sugar is released into the bloodstream more slowly. Complex carbohydrates also provide fiber, which is essential for overall and "down there" health. Examples of complex carbohydrates include whole grain breads, cereals, pasta, oats, brown rice, fruit, and vegetables.

Protein

Proteins are important nutrients; they are found in every cell in the body. The human body contains at least twenty thousand different proteins.[1] Protein is made up of subunits called amino acids. When we digest protein, it is broken down by enzymes (which are themselves proteins) into amino acids and small chains of amino acids called peptides. The amino acids and peptides are utilized by the body to make enzymes, hormones, neurotransmitters, DNA, and antibodies, along with forming muscles, bones, hair, skin, and nails (to name just a few structures). Protein also makes new cells and helps the body to repair damaged ones.

The recommended dietary allowance (RDA) of protein is 0.8 grams per kilogram of body weight.[2] This RDA can be considered the minimum amount of protein an adult needs on a daily basis. If you are active or are stressed by illness, you may need more protein in your diet. Those with kidney disease may require less protein because protein is eliminated by the kidneys. To determine your RDA for protein, multiply your weight in pounds by 0.36. For example, a sedentary man who weighs 180 pounds will need a minimum of sixty-five grams of protein daily. Unfortunately, there is a wide range of opinions on the "ideal" amount of protein. The National Institute of Medicine recommends that Americans get 10 to 35 percent of daily calories from protein.

Protein is found in meat, seafood, poultry, eggs, beans and peas, nuts and seeds, and dairy and soy products. Like carbohydrates, not all proteins are healthy choices. Red meat and processed meats (like pepperoni, salami, bologna) are high in saturated fat and sodium and are linked to cardiovascular disease, diabetes, and cancer. Healthier protein sources include seafood, white meat poultry with the skin removed, eggs, low-fat dairy products, and vegetarian options such as soy, beans, seeds, and nuts.

Fats

About twenty years ago, low-fat or no-fat diets were popular. Baked goods without fat were flying off the supermarket shelves. However, it didn't take long for most people to realize that they weren't losing weight or becoming healthier on these types of foods. Most of the fats in these products were replaced by sugar. Fats are actually an essential part of a healthy diet. These nutrients provide energy, assist in absorption of some (fat soluble) vitamins, and maintain healthy hair and skin, among other functions.

Like protein and carbohydrates, some fats are healthier than others. Saturated and trans-fats have been shown to raise bad (LDL) cholesterol and lower good (HDL) cholesterol. The American Heart Association recommends limiting saturated fat found in meat and some dairy products to under 6 percent of total calories.

Trans-fat is considered to be the unhealthiest type of fat. Trans-fat is manufactured by adding hydrogen to vegetable oil, which causes the oil to become solid at room temperature. Trans-fat, also known as partially hydrogenated oil, allows processed foods such as donuts, cookies, frosting, margarine, chips, and fried foods to have a longer shelf life. Replace saturated and trans-fats with the heart-healthy unsaturated fats.

Unsaturated fats include polyunsaturated fatty acids and monounsaturated fats. Healthy fats can lower bad (LDL) cholesterol and raise good (HDL) cholesterol levels. Polyunsaturated fats are found mostly in vegetable oils. Omega-3 fatty acids are a type of polyunsaturated fat that is found in fatty fish such as salmon, mackerel, and trout and also in walnuts and flaxseed. Monounsaturated fats contain vitamin E (an antioxidant) and can also reduce the risk of heart disease. This type of healthy fat is found in olives, nuts, avocados, sesame and pumpkin seeds, and olive, canola, and peanut oil.

Dietary Supplements: Is There a Pill for That?

More than 70 percent of Americans take dietary supplements, and we spend more than $30 billion a year on these products.[3,4] More than half of these consumers take the products for overall health, and about a third of consumers use the products to fill a perceived nutritional gap in their diet.

A well-balanced diet with the components described above is the best way to obtain all the nutrients necessary to stay healthy. However, some people may get a benefit from using supplements. Dietary or nutritional supplements include such ingredients as vitamins, minerals, herbal products, enzymes, and amino acids.

In this section, we'll discuss what dietary supplements may be beneficial for men for overall health as well as "down there" health. In addition, we'll cover some products that should be avoided. It's important to check with your health care provider before adding a supplement to your regimen. Some of these products may be harmful to you or may interfere with your other medications.

Supplements for Men's Overall Health

The following supplements may be beneficial for men (and women) for general health and well-being:

Multivitamins

Experts disagree (as they tend to do in the scientific community) on the value of taking a daily multivitamin. Most studies of people who took multivitamins failed to show a benefit. A relatively large study examined six thousand men over the age of sixty-five who took either a daily multivitamin or a placebo for over ten years. There was no benefit on memory loss for those who took the vitamin. Another study examined 1,700 people who had a previous heart attack. After about five years, those taking a multivitamin did not have a reduced risk of a second heart attack. Two studies showed that there may a small reduction in cancer risk from use of a daily multivitamin. An additional study showed that multivitamins may lower the risk of developing cataracts.[5]

So, what's the bottom line on whether we should be taking a daily multivitamin? There is no evidence that supports use of a multivitamin for everyone, but some men should consider it:

- Those who don't eat an optimal diet for any reason;
- Those who are strict vegetarians who may not be getting enough vitamin B12, iron, calcium, and zinc in their diet;
- Those who have restricted diets for weight loss or other reasons;
- Those who are recovering from surgery or have a serious illness that prevents normal digestion of nutrients;

- Those on medications that impair normal digestion of nutrients. For example, proton pump inhibitors and H2 blockers may prevent B12, vitamin C, calcium, magnesium, and iron from being digested.

Vitamin D

Up to 77 percent of Americans have a deficiency of vitamin D.[6] Vitamin D is a fat-soluble vitamin and is nicknamed "the sunshine vitamin" because it is produced by the skin in response to direct exposure of the sun's ultraviolet rays. Vitamin D is also found in foods such as salmon, sardines, egg yolks, and shrimp. Other foods (such as milk, cereal, and orange juice) may be fortified with vitamin D.

Our bodies synthesize vitamin D, and we also eat foods that contain vitamin D, so why do so many Americans have low vitamin D levels? Many of us avoid the sun and wear sunscreen to protect our skin. Correctly applied sunscreen can decrease vitamin D production by more than 90 percent.[7] In addition, most of us don't get enough vitamin D from the foods that we eat.

Vitamin D is an important nutrient for both men and women for several reasons. It regulates the absorption of calcium and phosphorus which are needed for bone growth and strengthening (see Chapter 9 on osteoporosis). Vitamin D plays a role in proper functioning of our immune system. In addition, it has been shown to reduce the risk of heart disease, multiple sclerosis, diabetes, influenza, and COVID-19. Vitamin D may prevent prostate, colon, and other types of cancers as well.

Because it is difficult to get enough vitamin D naturally, supplements do make sense for many adults. Check with your health care provider to make sure that taking vitamin D is beneficial for you. Current guidelines recommend 600 international units (IU) per day for those under the age of seventy and 800 IU daily for older men. Many experts recommend 800 IU to 1,000 IU for most adults, and doses up to 4000 IU daily are considered safe. Vitamin D supplements are available in two forms: vitamin D2 (ergocalciferol) and vitamin D3 (cholecalciferol). The D3 form is considered to be more active and is the preferred supplement.

Calcium

As discussed in Chapter 9, calcium is vital for bone health and prevention of osteoporosis. Most men should consume a total of 1,000 to 1,250 mg of calcium daily, depending on age. Dietary sources of calcium are best and include dairy products, leafy green vegetables, bony fish, and fortified food and drinks. Men who don't get enough calcium in their diet should consider taking a calcium supplement if recommended by their health care provider. However, too much calcium may be harmful. Some research has linked excess calcium from supplements to kidney stones, prostate cancer, and heart disease, but more evidence is needed. Calcium citrate may be the preferred type of calcium because it is absorbed by the body better than other types of calcium supplements.

Fish Oil (Omega-3 Fatty Acids)

Fish and fish oil are sources of omega-3 fatty acids, which are important in the formation of every cell in the human body. Omega-3 fats are known as "essential fats" because they can't be produced in the body; we must get these nutrients from food. The two most important omega-3 fatty acids are eicosatetraenoic acid (EPA) and docosahexaenoic acid (DHA). These fats support heart, brain, and nerve cell function and assist in healing of tissues. They also have anti-inflammatory properties.

Because omega-3 fats are essential, fish oil has been promoted as a miracle prevention and cure for many disorders. In fact, this supplement is used daily by almost 20 million Americans—up to 10 percent of us.[8] The benefits of fish oil are controversial, but they lower triglyceride levels and protect against heart disease and stroke, especially in those who have already been diagnosed with these diseases.[9] Omega-3 fats may also be helpful for joint and eye health. Other potential uses such as prevention or treatment of depression, diabetes, cancer, memory loss, and ADHD have not been supported by large-scale studies. One study showed a potential correlation between higher levels of omega-3 fatty acids and prostate cancer, but this study has been criticized by many experts.

The best way to ensure an adequate intake of omega-3 fatty acids is to eat fish at least twice weekly. Good sources of omega-3s include trout, salmon, sardines, herring, tuna, and mackerel. Men who don't eat the recommended amount of fish and those with known heart disease should ask their health care provider about fish oil supplements. These products do have some potential side effects such as belching, nausea, bad breath, and heartburn. Side effects can be reduced by taking these supplements with food. In addition, large doses of fish oil can increase your risk of bleeding, especially if you have a bleeding disorder, bruise easily, or are taking a medication that thins your blood such as Coumadin or a nonsteroidal anti-inflammatory (NSAID).

Zinc

Zinc is beneficial throughout the body. It is a mineral found in every cell and it helps us with metabolism, immunity, and wound healing. In addition, zinc is known to help men with sexual health. Zinc impacts testosterone levels and might even be used as a treatment for erectile dysfunction. The recommended daily allowance of zinc is 11mg for men. Good dietary sources include oysters and other shellfish, meat, legumes, eggs, nuts, and seeds.

Vitamin E

Vitamin E is an antioxidant that protects the body from damage caused by harmful chemicals called free radicals. Vitamin E might also help prevent heart disease, cancer, and age-related macular degeneration. Most people get enough vitamin E in their diet by eating foods such as nuts, seeds, and olive oil.

Magnesium

Magnesium is a mineral that is critical for many processes in the body. It helps to keep the brain, heart, and muscles functioning properly and may lower blood pressure, improve sleep quality, lower blood sugar, and reduce inflammation. Magnesium might also help people regulate mood. Although magnesium is important, almost half of Americans are deficient in their intake.[10] Foods high in magnesium include nuts, seeds, leafy greens, whole grains, legumes, and beans. You may be happy to hear that chocolate also contains magnesium.

Resveratrol

Resveratrol is a plant compound that acts as an antioxidant in the body and may help prevent cancer and heart disease. It has been shown to lower blood pressure, cholesterol levels, and blood sugar and can improve brain function. Resveratrol may slow down the aging process. It is found naturally in grapes and red wine and can be consumed as a supplement.

Vitamin B3 Analogs

Niacinamide (NMN) is a form of vitamin B that is similar to niacin but is associated with less flushing. Niacinamide is a precursor of NAD+, a coenzyme needed for production of energy in all cells in the body. Depletion of NAD+ decreases energy production in cells and increases oxidative stress, damage to DNA, mental decline, and inflammation. The supplement form of niacinamide (NMN) may slow down the aging process in the body.

Supplements for "Down There" Health

The following section discusses supplements used for two common disorders in men.

Benign Prostatic Hyperplasia (BPH)

Benign prostatic hyperplasia, or an enlarged prostate gland, is common in men as they age. Symptoms of BPH include frequent urination, waking at night to urinate, a sudden urge for urination, weak stream, and dribbling. Several types of medications are FDA-approved for treatment of BPH, and many men opt for surgery to relieve symptoms. Dietary supplements are also marketed for treatment of symptoms of BPH. The American Urological Association (AUA) does not support the use of these products because of a lack of evidence of effectiveness and possible lack of safety. The most commonly used supplements are listed below:

- Saw Palmetto

- Pygeum africanum
- Beta-sitosterol
- Rye grass pollen

Erectile Dysfunction (ED)

Erectile dysfunction is the inability to obtain or maintain an erection. The incidence of erectile dysfunction increases with age, and this disorder negatively affects quality of life for those who suffer from it. "Failure to launch" can happen to anyone from time to time, but chronic ED may be caused by physical problems like diabetes, heart disease, high blood pressure, or low testosterone levels; emotional problems like depression or stress; or substances such as alcohol, medications, illicit drugs, or cigarettes.

The most important piece of treating ED is to find out the potential cause. Fortunately, ED may be treated effectively once the cause is known. All of us know about the little blue pill and similar medications, but do dietary supplements work just as well? Natural products have been used in African, Chinese, and other cultures for centuries. However, these products haven't been well studied for the treatment of ED. Safety may also be a concern. Notify your health care provider before taking any supplements for ED. As with drugs for BPH, the American Urological Association (AUA) does not support the use of these products. Some of these products are discussed below:

L-Arginine

L-arginine is an amino acid found in meat, fish, and poultry which causes increased blood flow in the body. This supplement is considered to be possibly effective for the treatment of

ED. Side effects are uncommon but can include nausea, diarrhea, and worsening of asthma symptoms.

Dehydroepiandrosterone (DHEA)

Dehydroepiandrosterone is a natural hormone that is used by the body to produce testosterone. Like testosterone, DHEA levels decrease as men age. Some evidence shows that DHEA may be effective for ED. This supplement may cause acne but is otherwise safe. Do not use this product if you have prostate cancer.

Ginseng

Ginseng has been used for sexual performance in the Chinese culture for centuries. It works to promote relaxation of smooth muscle in the penis and also increases levels of the neurotransmitter dopamine in the brain. Small studies suggest that ginseng may be effective for ED, but larger trials are needed. This supplement may cause nervousness, insomnia, headaches, dizziness, or upset stomach. It also might interfere with drugs for diabetes and blood thinners.

Horny Goat Weed (Epimedium)

This supplement gets the prize for the best name! It is another herb that has been used by the Chinese for many years to assist with sexual performance. Although studies in rats (not goats) look promising, no studies have been performed in humans.

Supplements to Avoid: "Buyer Beware"

Most of us assume that natural vitamins, minerals, and supplements are beneficial to our health and provide nutrients that we may be missing in our diet. We also assume that products available at our local health food store are safe at the very least. Unfortunately, this may not be the case. The Federal Drug Administration (FDA) reviews all prescription and over-the-counter drug products for efficacy and safety. However, dietary supplements are regulated differently. Manufacturers of dietary supplements are not required to obtain FDA approval before marketing a nutritional supplement. The manufacturers themselves are responsible for the safety of their product. As a result, these products are considered safe until they are proven to be unsafe.

So, what's the bottom line on supplements? Most dietary supplements are safe. However, you should take some steps to make sure that you're getting a good quality product. First, look for a quality seal from one of the four organizations that performs testing and inspections of supplements and the plants where they are manufactured. These include Consumer Lab, Natural Products Association, NSF International, and US Pharmacopeia. Supplements that display a seal from one of these companies contain the ingredients listed on the label, are manufactured properly, and don't include toxic contaminants. You may also want to contact the manufacturer to ask about any research that has been done on the product and reported side effects. Finally, check the FDA website to make sure that the supplement has not been recalled. (https://www.fda.gov /safety/recalls-market-withdrawals-safety-alerts)

Consumer Reports has published a list of supplement ingredients to avoid.[11] These substances are potentially harmful depending on a person's medical history, amount of ingredient contained in the product, and the length of time the substance is taken. Let's review the fifteen harmful ingredients listed by *Consumer Reports*:

1. Aconite (Wolfsbane, Monkshood)

This product is used to reduce inflammation and joint pain. It has also been used to treat heart failure. However, aconite contains a fast-acting poison that causes serious side effects such as nausea, vomiting, paralysis, weakness, breathing problems, and possibly death. Aconite should never be taken by mouth or applied to the skin.

2. Bitter Orange (Methylsynephrine)

Bitter orange is made from the fruit, leaf, and peel of a citrus plant. The supplement is a stimulant that is used as an appetite suppressant for weight loss and for various other disorders including upset stomach, nasal congestion, diabetes, and chronic fatigue. It is available as an oil and can be applied to the skin for fungal infections like athlete's foot and jock itch or inhaled as an aromatic oil.

Bitter orange is also found in marmalades and liquors and in cosmetics and soaps. The herb is probably safe in the amounts found in food products and for topical use on the skin. However, the supplement can cause serious side effects: high blood pressure, cardiac arrest, heart rhythm abnormalities, and headaches. Bitter orange is a stimulant and can interact with other drugs that excite the central nervous system.

3. Caffeine Powder (1,3,7-trimethylxanthine)

Caffeine seems like a safe substance since many of us drink coffee, tea, and soda which contain caffeine. This supplement is used to improve mental alertness and attention, to lose weight, and for various medical conditions such as pain, asthma, and migraines. Caffeine is probably safe for most adults when used appropriately. However, this product can be harmful in high doses. It can cause insomnia, nervousness, nausea and vomiting, stomach irritation, anxiety, chest pain, ringing in the ears, seizures, heart arrhythmia, cardiac arrest, and possibly death. Caffeine also interacts with other stimulant drugs.

4. Chaparral (Creosote Bush, Greasewood)

Chaparral is a flowering plant found in the desert. It has been used by Native Americans for infections and diseases such as tuberculosis, sexually transmitted diseases, and snakebites. It has also been used for weight loss, cancer, skin rashes, and even the common cold. However, this supplement has been associated with liver and kidney damage. It should be avoided.

5. Coltsfoot (Coughwort, British Tobacco)

Coltsfoot is also a flowering plant; it's in the daisy family and it resembles dandelion. It is found in Europe and Asia. This supplement has been used for respiratory problems such as bronchitis, asthma, cough, and sore throat. Coltsfoot contains toxic substances called pyrrolizidine alkaloids (PAs), which have been linked to liver damage and possibly cancer. Coltsfoot is unsafe and should be avoided.

6. Comfrey (Blackroot, Slippery Root, Blackwort)

Comfrey is a plant with a black root, large leaves, and small purple- or cream-colored flowers. Native to Europe, comfrey has been used as a tea for all sorts of ailments including ulcers, heavy menstrual periods, diarrhea, cough, chest pain, cancer, and sore throat. It has also been applied to the skin for bruises, joint inflammation, and wounds. Comfrey also contains pyrrolizidine alkaloids, which, as noted earlier, can cause liver damage and possibly cancer.

7. Germander (Teucrium Chamaedrys)

This plant is found in the Mediterranean. It is used for fever, as a digestive aid, for weight loss, and to freshen breath. The use of germander is unsafe, possibly leading to liver damage and hepatitis.

8. Greater Celandine

Greater celandine is a perennial plant in the poppy family. This herb is used for various digestive problems such as upset stomach, constipation, and irritable bowel syndrome. It can also be applied to the skin to treat rashes and warts. The plant contains several toxic alkaloids and may be associated with liver damage.

9. Green Tea Extract Powder (Camellia Sinensis)

This one is a little tricky. Green tea contains polyphenols, including one called catechin, which works as an antioxidant. Green tea has been used for weight loss, cancer prevention, alertness, depression, headaches, and Parkinson's disease, just to name a few. It has also been used topically for genital warts and to soothe sunburn. Green tea is probably effective for genital warts and to lower cholesterol. It is possibly effective for several other conditions, including the prevention of high blood pressure and some cancers. Green tea is also considered to be safe when consumed as a drink in moderate doses.

Nevertheless, green tea contains caffeine and in high doses can lead to nervousness and anxiety, sleep problems, vomiting, diarrhea, irritability, irregular heartbeat, tremor, heartburn, dizziness, ringing in the ears, seizures, confusion, and liver damage. Green tea may also reduce the absorption of iron from food. Consuming very high doses of green tea might be fatal.

10. Kava (Kava Kava, Piper Methysticum)

Kava is a drink or extract made from the leaves of the piper methysticum plant native to the Western Pacific islands. Kava is a popular drink in the Pacific used to calm anxiety and stress and to treat insomnia. It has also been used for attention deficit disorders, epilepsy, depression, headaches, and pain. Use of kava has been linked to liver damage and death.

11. Lobelia (Asthma Weed, Vomit Wort)

Lobelia is a type of flowering plant found in warm climates. It is used for breathing problems such as cough, asthma, bronchitis, and shortness of breath in newborns. It can also be used topically for muscle pain, bruises, insect bites, and poison ivy. This herb is associated with nausea, vomiting, diarrhea, dizziness, tremors, confusion, coma, and possibly death.

12. Pennyroyal Oil (Mentha Pulegium)

Pennyroyal is a type of flowering plant that has been used for centuries to eliminate pests and to induce abortion. In addition to the uses mentioned, this herb has been used for cold relief, cough, fever, headache, and indigestion. Pennyroyal is toxic and has caused liver and kidney failure, seizures, and death.

13. Red Yeast Rice (Monascus Purpureus)

This product is made from yeast that is grown on rice. Red Yeast Rice is another of the "mixed bag" supplements. It is effective for lowering cholesterol because it contains a substance called monocolin, which is similar or identical to lovastatin (an FDA-approved drug for the treatment of high cholesterol). The FDA banned red yeast rice for several years, but products without monocolin have reappeared in the last ten years. This product may contain citrinin, a poison that may cause kidney damage. Like "statin" cholesterol-lowering drugs, red yeast rice may also cause liver and muscle damage. This supplement should not be taken with statin drugs.

14. Usnic Acid (Beard Moss, Tree Moss)

Usnic acid is a lichen that grows on trees. A lichen is a combination of algae and fungus that grow together for a mutual benefit. This supplement can be used for weight loss, fever, pain relief, wound healing, and as a cough expectorant. It can also be used for inflammation of the throat and mouth. Usnic acid is associated with liver damage and should be avoided.

15. Yohimbine

Yohimbine is used to treat erectile dysfunction and low libido. It has also been used to treat anxiety, depression, and obesity. Yohimbine can cause serious side effects such as high blood pressure, seizures, rapid heart rate and rhythm disturbances, kidney failure, heart attack, and possibly death.

Whatever Happened to Sam?

Sam's initial body mass index was 41.1, which put him into the "obese" category. Sam was rightfully concerned about his diet. The types of foods that he had been eating are high in fat, refined carbohydrates, and salt. Sam suffers from high blood pressure, but his blood pressure might be better controlled if Sam lost a few pounds. Caffeine powder is not recommended in people with high blood pressure.

Sam was motivated to lose weight. He was able to lose forty pounds in a year by bringing healthy meals with him on the road. Sam also made better choices when eating out. He traded in his burgers and pizza for salads with lean protein. He also limited his alcohol consumption. Sam's blood pressure has decreased, and he no longer needs medication.

Bottom Line

According to the book *Dietary Guidelines for Americans 2015–2020*, about half of adults in the US have at least one preventable chronic disease. Many of these diseases are related to poor eating practices and lack of physical activity. Also, more than two thirds of adults (and almost one third of children) are overweight or obese. The bottom line is that nutrition is a huge factor in our health. As mentioned above, a healthy eating pattern includes whole fruits, a variety of vegetables, whole grains, low-fat dairy products, healthy oils, and a variety of lean protein. In addition, we should limit saturated fat, trans-fat, sodium, and sugar. Supplements may be beneficial for some individuals, but you should check with your health care provider before taking any supplements.

References

1. Ponomarenko, Elena A., Ekaterina V. Poverennaya, Ekaterina V. Ilgisonis, Mikhail A. Pyatnitskiy, Arthur T. Kopylov, Victor G. Zgoda, Andrey V. Lisitsa, and Alexander I. Archakov. "The Size of the Human Proteome: The Width and Depth." *International Journal of Analytical Chemistry* (May 19, 2016): https://doi.org/10.1155/2016/7436849.
2. Gunnars, Kris. "Protein Intake—How Much Protein Should You Eat per Day?" Healthline, October 2, 2020. https://www.healthline.com/nutrition/how-much-protein-per-day.
3. Council for Responsible Nutrition. "Dietary Supplement Use Reaches All Time High." News release, September 30,

2019. https://www.crnusa.org/newsroom/dietary-supplement-use-reaches-all-time-high.

4. Haspel, Tamar. "Most Dietary Supplements Don't Do Anything. Why Do We Spend $35 Billion a Year on Them?" *Washington Post,* January 27, 2020. https://www.washingtonpost.com/lifestyle/food/most-dietary-supplements-dont-do-anything-why-do-we-spend-35-billion-a-year-on-them/2020/01/24/947d2970-3d62-11ea-baca-eb7ace0a3455_story.html.

5. Council for Responsible Nutrition. "Supplement Use Among Younger Adult Generations Contributes to Boost in Overall Usage in 2016—More than 170 Million Americans Take Dietary Supplements." October 27, 2016. https://www.crnusa.org/newsroom/supplement-use-among-younger-adult-generations-contributes-boost-overall-usage-2016-more.

6. Lefevre, Michael L. "Screening for Vitamin D Deficiency in Adults: U.S. Preventative Services Task Force Recommendation Statement." *Annals of Internal Medicine* 162, no. 2 (2015): 133. https://doi.org/10.7326/m14-2450.

7. The Nutrition Source. "Vitamin D." May 26, 2015. https://www.hsph.harvard.edu/nutritionsource/vitamin-d/.

8. National Center for Complementary and Integrative Health. "7.8% of U.S. adults (18.8 million) used Fish oil/Omega-3 fatty acids." National Institutes of Health, August 11, 2016. https://nccih.nih.gov/research/statistics/NHIS/2012/natural-products/omega3 (article removed) .

9. O'Connor, Anahad. "Fish Oil Claims Not Supported by Research." *Well* (blog). *New York Times,* March 30, 2015. https://well.blogs.nytimes.com/2015/03/30/fish-oil-claims-not-supported-by-research/.

10. Workinger, Jayme L., Robert P. Doyle, and Jonathan Bortz. "Challenges in the Diagnosis of Magnesium Status." *Nutrients* 10, no. 9 (September 2018): 1202. https://doi.org/10.3390/nu10091202.

11. Gill, Lisa. L. "10 Supplements to Always Avoid. Consumer Reports, last modified December 8, 2022. Accessed May 2, 2022. https://www.consumerreports.org/vitamins-supplements/15-supplement-ingredients-to-always-avoid-a1185290021/.

CHAPTER 15

How to Select a Physician

Francois visited his primary care doctor for his annual checkup; surprisingly, a screening PSA test indicated an elevated level. Francois was told by his primary care doctor that he had to see a urologist for further evaluation, as his elevated PSA level could be a marker for prostate cancer. Francois was shocked; he told the doctor he had no urinary symptoms or pain in his "nether" regions. His doctor provided him with a referral to a urologist.

Finding a Primary Care Doctor

One of the most important decisions you will make regarding your health is to find a doctor who will care for your health care needs. The right decision can simplify your life and possibly keep you healthy and well. The wrong decision can ruin your life and even make it shorter. This chapter will provide you with suggestions that will help you make the right decision and lead you to a doctor who will be the best fit. We will also discuss how to find a urologic surgeon if the need arises.

The search begins with finding a doctor who will meet your medical needs. If you have a medical condition, you will want to find a doctor who is comfortable treating that condition or has experience with your preexisting medical illness.

There are several different types of doctors that are identified as primary care physicians: family practice, internal medicine, and geriatrics, or general practice.

Family Practice

Family practice physicians treat patients of all ages. They are generalists who can treat a wide variety of conditions. Many can also treat ailments you'd normally see a specialist for, like sports injuries or mild allergies.

This category is one area where you might also find osteopaths, physicians who practice a type of alternative medicine with a special focus on the musculoskeletal system. Osteopaths are identified by the "DO" initials (Doctor of Osteopathic Medicine) after their name instead of "MD" They are well-trained and fully licensed doctors, and their treatment and skills are going to be very similar to MDs.

Internal Medicine

Internal medicine physicians typically treat adults and specialize in the prevention, diagnosis, and management of disease and chronic conditions. They have several years of additional training beyond medical school.

Geriatrics

Geriatricians are medical doctors who specialize in caring for elderly patients. If you are elderly, start your search by getting a reference from someone you know and trust. Ask friends, family members, neighbors, or coworkers if they have a doctor they like. If your doctor is retiring, ask your current physicians for a recommendation. They know your medical condition, your personality, and your needs, and they will likely be able to suggest a physician who will be a good fit.

Insurance Issues

If you have health insurance, you may need to choose from a list of doctors in your plan's network (doctors who take your insurance plan). Some insurance plans may let you choose a doctor outside your network, but going outside the network usually requires you to pay more of the cost.

Another method is to go to the insurance company's website to search for a doctor. If you find a doctor from the website that accepts your insurance, then call the doctor's office and ask their representative to confirm that they take your plan. If you don't have health insurance, you'll have to pay for health care out of pocket (on your own.)

Now that you have found a listing of doctors who will accept your insurance, make a list of the doctors you're interested in. Be sure to think about how easy or difficult it will be to travel to their office locales. Then call each office to learn more about their operation. The answers to the following questions may help you make the best decision.

Questions About the Doctor
- Is the doctor taking new patients?
- Is the doctor part of a group practice? If so, who are the other doctors that might help care for me?
- Who will see me if my doctor isn't available?

- Which hospital(s) does the doctor use?
- Does the doctor have experience treating my medical conditions?
- Is the doctor board certified?
- Does the doctor have any special training or other certifications?

Questions About the Office

- Are evening or weekend appointments available? What about virtual appointments over the phone or on a computer (telemedicine)?
- What is the cancellation policy?
- How long will it take to get an appointment?
- How long do appointments usually last?
- Can I get lab work and X-rays done in the office?
- Is there a doctor or nurse who speaks my preferred language?
- Is there parking in the doctor's office building or nearby?

We also suggest that you make an appointment with the potential choice of a primary care doctor and ask to meet the doctor. This functions as an interview, and most doctors will accept such an appointment; expect to be charged for the doctor's time. Consider asking the doctor questions about their level of availability and the access you will have for routine appointments, urgencies, and emergencies. You will want to know how you will receive lab and imaging reports. Will these results be sent to you by email, use of the practice portal, or a phone call from the physician?

Many patients are interested in telemedicine or having virtual visits. Ask your potential primary care doctor if that service is available. You will want to know about calling the doctor for emergencies after the practice closes and on weekends and holidays. You will also want to know who takes calls for the doctor when they are not available. You might then ask if the doctor would share the names of any of the patients who might allow an interview. Of course, this can only be done if the doctor's existing patients agree and give permission to take a call from a potential new patient.

After this interview with the doctor and their staff, you should be able to answer the following questions. Did the doctor and office staff:

- make me feel comfortable during my appointment?
- explain things in a way that was easy to understand?
- listen carefully to me?
- show respect for what I had to say?
- know important information about my medical history?
- spend enough time with me?
- give me a chance to ask questions?

If you answer "no" to any of these questions, you may want to keep looking.

Finding a Urologic Surgeon

A urologist specializes in diseases of the kidney, bladder, prostate gland, and testicles. The urologist usually completes four to five years of training after medical school. Often the urologist will take a one- or two-year fellowship, which is additional training to specialize in a specific aspect of urology such as urologic cancer, male infertility, or urinary incontinence.

Prostate cancer surgery is a very difficult operation. It takes not only skill, but the kind of expertise and training that only comes from involvement in several hundred procedures. For several years the urologic surgeon is a doctor in training, learning from the sidelines before demonstrating the skills needed to operate meticulously under the guidance of an expert surgeon.

The very best prostate surgeons specialize in the prostate. That's often all they do, and they perform many of these procedures, sometimes more than two hundred cases each year.

Because there are marginally trained and barely competent surgeons out there, you can't trust everything you read on the internet or from hospital propaganda posted on their websites.

As for you, well, this is your one shot at this. Do your due diligence. Please consider the following before you go under the knife:

First, select a high-volume center that does a lot of prostate cancer surgeries. If they do a lot of these procedures, they will have a team in place, and as a result, everyone is going to be better at helping you with your recovery after surgery. Prostate cancer surgery patients have unique needs that are different from other surgical procedures. If the hospital has a team trained to care for men after prostate surgery, the nurses will know how to take care of recovering prostate surgery patients. Some large centers even have a floor or a wing just for men with prostate cancer—and not for other postoperative patients, whose postop needs are very different from men who have had prostate surgery.

Two websites can direct you to a high-volume center. The National Cancer Institute's website designates cutting-edge cancer treatments for patients in communities across the United

States (http://www.cancer.gov/research/nci-role/cancer-centers /find).

The National Comprehensive Cancer Network is another helpful site that identifies topnotch cancer centers (https:// www.nccn.org/).

Prostate cancer is a complicated medical condition, and there is no "one size fits all" for every patient. The treatment that your friend or one of his family members had might not apply to your condition. Therefore, you will probably want to obtain advice from several specialists. If you go to a center that has specialists who focus on the treatment of prostate cancer, you will receive the opinion of a team of experts, not just one, and the benefit is a more thorough and thoughtful approach to your treatment.

Seek a practice or hospital where different specialties work together. Top centers have teams in place that include experts from different specialties: urology, radiation oncology, medical oncology, and pathology all working together on prostate cancer. Depending on a patient's diagnosis, age, and other medical conditions, he may be a perfect candidate for surgery, for radiation therapy, or for simply watchful waiting (see Chapter 3).

You want to ask your potential surgeon about their results. You might begin the conversation by asking how many surgical procedures they have done and how many procedures they do each year. Most surgeons keep cumulative records and can provide you with their statistics if you ask. The surgeons might also publish their results in peer-reviewed literature, and you might ask for copies of those papers or reports. Ask the surgeon these important questions:

- Was the PSA level undetectable following surgery?

- What percentage experienced erectile dysfunction (impotence) after surgery?
- What percentage experienced urinary incontinence following surgery?
- What was the average duration of postsurgical impotence and/or incontinence?

Merely asking these questions demonstrates to the surgeon that you are knowledgeable and informed and that you and are going to be using this information to help make your decision.

The most important factors are the reputation of the institution, the surgical department, and the surgeon. Consider asking the surgeon to provide you with the names of their patients who have agreed to speak to other patients about their hospital experience. This will validate that the surgeon has happy and grateful patients; that he cares enough to compile such data; and that he appreciates the importance of a large support network. Most surgeons have satisfied patients who are willing to share their surgical experience with potential patients. We suggest preparing questions that you would like to ask potential patients once you have opportunity to speak to them in person, by Zoom, or on the telephone.

We also suggest that you ask more than one doctor to recommend the best prostate surgeon(s) available. Some doctors practice in groups and will recommend a specialist in their group. Therefore, it's a good idea to ask different doctors in different practices.

It is very common for men to do an online review to find more information about prostate cancer. Be wary of the glitzy reviews or ads on the internet. For the most part, online reviews are unreliable; we suggest that you discount them.

Research has shown poor correlations between online reviews and outcomes.

Speaking to other patients and local doctors is a much better idea. We also recommend that you check with prostate cancer support groups in your area and ask the men about their own experiences and advice on a surgeon. The internet is full of false accusations and glamorization of surgeons and their hospitals and departments. Hospital websites sometimes advertise benefits and programs that are often not present or real. We realize that seems to be a sinister characterization, but it's a pervasive reality in today's society and within the medical profession.

And finally, don't worry about offending the doctor with questions or by getting a second opinion. Patients ask for second opinions all the time. You are paying the doctor, not the other way around. (NOTE: This doesn't mean you should be rude or disrespectful; it just means you shouldn't feel intimidated or believe you are being a bad guy simply for doing your homework.) If the situation were reversed, do you think your doctor would not make every effort to find the best possible surgeon? It's your prostate, your recovery, and your life. You don't want to be one of those guys saying afterward, "My surgeon was not very good."

If you have been diagnosed with cancer, finding a doctor and a treatment facility for your cancer care are the critical first steps to getting the best treatment possible.

Back to Francois!

Francois was referred by his primary care doctor to a urologist. The urologist recommended either of two options: surgery to remove the prostate or radiation therapy. Francois's urologist referred him to a radiation oncologist and a urologic surgeon who specialized in robotic prostate gland removal. Francois met with both doctors and opted for surgery. He had the surgery, and his PSA is now undetectable. He follows up with his urologist for symptom checks and PSA tests every three months.

Bottom Line

You will have many things to consider when choosing a doctor. It's important for you to feel comfortable with the doctor and the specialist that you choose because you will be working closely with that person to make decisions about your health care.

CHAPTER 16

Exercise and Stretching

Boyd is a slightly overweight middle-aged man with diabetes, high blood pressure, high cholesterol levels, and erectile dysfunction. He had a visit with a urologist who said, "Boyd, would you like me to give you a prescription for medication that will lower your blood pressure, decrease your need for medication for diabetes and your elevated cholesterol level, promote weight loss, decrease your risk of heart disease, improve your mood, increase your libido, decrease your joint pain, decrease your risk of cancer, improve your sleep, enhance your immunity, and make your penis one to two inches longer?" Boyd responds with extreme interest in finding out more about this miracle medication. The doctor says, "Boyd, it's not a medication; it's exercise!"

Have you ever wondered about one magical activity that could reduce the brain fog that comes with age, prevents depression and stress, lowers blood pressure, and lowers the odds of having heart problems? There is an abundance of evidence that one such activity that covers it all and more is exercise. Exercise is the physical activity intended to improve or maintain physical health. This chapter will review the benefits of exercise and how to initiate a physical exercise program to enhance your health.

Benefits of Physical Activity

Regular physical activity is one of the most important things you can do for your health. Physical activity can improve your brain health, help control your weight, reduce chronic diseases, strengthen bones and muscles, decrease your risk of cancer, and possibly stave off dementia. Very few lifestyle choices have as significant an impact on your health as regular physical activity. Almost anyone can engage in and benefit from physical activity regardless of age, ability, ethnicity, shape, or size.

Immediate Benefits

Some benefits of physical activity happen immediately after moderate-to-vigorous physical activity. Benefits include improved thinking and reduced short-term feelings of anxiety. Regular physical activity can help keep your thinking, learning, and judgment skills sharp as you age. It can also reduce your risk of depression and anxiety and help you sleep better.

Weight Management

Weight control follows the laws of thermodynamics. If you simply consume more calories than you burn through daily activities, you gain weight. And the reverse also applies: burn more calories than you consume, and you will lose weight. *To lose weight, you need to move more or eat and drink less.* Getting to and staying at a healthy weight requires both regular physical activity and healthy eating.

Prevention of Cardiovascular Disease

Heart disease and stroke are two leading causes of death in the United States. Regular physical activity lowers your risk for these diseases and can lower your blood pressure and improve your cholesterol levels.

Prevention of Type 2 Diabetes

Regular physical activity can reduce your risk of developing adult-onset diabetes mellitus or type 2 diabetes. A recent study showed that men who exercised at least three hours a week were 39 percent less likely to develop diabetes.[1]

Prevention of Some Cancers

Regular physical activity lowers your risk of developing several common cancers. Those who engage in greater amounts of physical activity have a reduced risk of developing cancers of these organs:

- Bladder
- Colon (proximal and distal)
- Esophagus (adenocarcinoma)
- Kidney
- Lung
- Stomach
- Prostate

- Bladder
- Breast (Yes, men do get breast cancer; see Chapter 13.)

If you are a cancer survivor, regular physical activity helps improve your quality of life. It may decrease the likelihood of recurrence of your cancer.

Strengthening of Bones and Muscles

As you age, it's essential to protect your bones, joints, and muscles. Keeping bones, joints, and muscles healthy ensures that you can do your daily activities and be physically active. Muscle-strengthening activities like lifting weights can help you increase or maintain your muscle mass and strength. This is especially important for older adults who experience reduced muscle mass and strength with aging. Slowly increasing the weight and number of repetitions you do as part of muscle-strengthening activities will give you even more benefits, no matter your age.

Safely Performing Daily Activities and Preventing Falls

Physically active middle-aged or older adults, like Boyd, have a lower risk of functional limitations than inactive people. For older adults, doing a variety of physical activities improves physical function. It decreases falls and fractures of the hips and spine. Hip fracture is a severe health condition that can result from a fall. Breaking a hip may have life-changing adverse effects, including increased mortality. Physically active people have a lower risk of hip fracture than inactive people.

Increased Longevity

More than one hundred thousand deaths per year could be prevented if US adults ages forty and older increased their moderate-to-vigorous physical activity by a small amount. Even ten minutes more a day makes a difference.[2]

Walking or taking more steps each day also helps lower the risk of premature death from all causes. For adults younger than sixty, the risk of sudden death leveled off at about eight thousand to ten thousand steps per day. For adults sixty and older, the risk of premature death leveled off at about six thousand to eight thousand steps per day.[3]

Numerous long-term studies have shown that regular exercise may prolong your life by as much as seven years.[4]

Managing Chronic Health Conditions & Disability

Regular physical activity can help people manage existing chronic conditions and disabilities. For example, regular physical activity can:

- Reduce pain and improve function, mood, and quality of life for adults with arthritis;
- Help control blood sugar levels and lower the risk of heart disease and nerve damage for people with type 2 diabetes;
- Help support daily living activities and independence for people with disabilities.
- Increase energy levels and endurance.

Inactive people—"couch potatoes"—often complain that they don't have any energy. Regular exercise increases energy levels.

That's the good news. The bad news is that it doesn't happen overnight. Just as the weightlifter won't develop a physique that looks like Arnold Schwarzenegger after one day at the gym, the energy benefits from exercise won't occur for several weeks or a month after engaging in a daily exercise program.

Endorphins and Exercise

Have you ever gone for a run or worked out and felt a sense of euphoria afterward? That heightened feeling of joy results from the release of endorphins after exercise. Endorphins are the chemicals in your brain that help you cope with pain, stress, and anxiety and help you feel good and even euphoric. Endorphins are neurotransmitters or hormones that are released in response to exercise. These neurotransmitters are also released by laughing, dancing, and driving a new car. Sexual intimacy is also characterized by a surge of endorphins.

The Psychological Benefit of Endorphins

Improved self-esteem is a key psychological benefit of regular physical activity. The stimulation of the release of endorphins interacts with the receptors in your brain that reduce your perception of pain and also trigger a positive feeling in the body, similar to that of morphine. The receptors endorphins bind to are the same ones that bind some pain medicines. However, unlike morphine, activating these receptors by the body's endorphins does not lead to addiction or dependence. For example, the feeling that follows a run or workout is often described as "euphoric." That feeling, commonly referred to as a "runner's high," can be accompanied by a favorable and energizing outlook on life. Endorphins also act as analgesics and diminish the perception of pain. They also act as sedatives, promote sleep, and serve as an antidote to insomnia.

Below is a list of some of the many benefits of endorphins:

- Reduced depression
- Improved self-esteem
- Reduced anxiety
- Reduced pain
- Boosted immunity
- Regulated appetite

Even though the release of endorphin varies among individuals (and often involves the release of other feel-good chemicals like dopamine, serotonin, and oxytocin), it's our body's way of protecting us from the harmful effects of pain and distress. It propels us forward towards feelings of pleasure, joy, and release. Without your endorphins, the world would seem dull and joyless.

Four Components of Fitness

When starting an exercise plan, most of us choose one type of activity. Research has shown that we should try to incorporate all four types of exercise. These include endurance, strength, balance, and flexibility.

Endurance Activities

These activities are the ones that increase your heart rate and breathing. Some examples of this type of exercise include individual and team sports like running, tennis, soccer, swimming, and fast-paced walking.

Strength or Resistance Training

This exercise type uses weights or resistance bands to increase muscle strength, important in providing us the power and mobility to perform daily tasks without injury. In addition, having strong muscles can prevent falls.

Balance Exercises

Balance exercises also help prevent falls. They help large and small muscles stabilize the body. Examples of these exercises include practicing yoga, standing on one foot, walking in a straight line, and Tai Chi.

Flexibility Activities

Finally, flexibility is an important aspect of fitness. Being flexible allows us to move without injury. Dynamic stretching (stretches with movement) allows us to prepare our body for exercise, and static stretching (active isolated stretches) helps us to prevent stiffness and injury after physical activity. Examples of dynamic stretching include hip and arm circles, and leg swings. Examples of static stretches include touching toes for thirty seconds or the runner's stretch.

Starting Your Exercise Program

You can start a fitness program in five steps (guidelines modified from Mayo Clinic: https://www.mayoclinic.org/healthy-lifestyle /fitness/in-depth/fitness/art-20048269).

1. Assess your fitness level and obtain your baseline physical activity.

Start by recording this data:

- Your pulse rate before and immediately after walking one mile (1.6 kilometers)
- How long it takes to walk one mile, or how long it takes to run 1.5 miles (2.41 kilometers)
- How many push-ups you can do at one time
- How far you can reach forward towards your toes while seated on the floor
- Your waist circumference, just above your hipbones
- Your body mass index (height in meters divided by weight in kilograms; a free BMI calculator is available at https://www.nhlbi.nih.gov/health/educational/lose_wt/BMI/bmicalc.htm)

2. Design your fitness program.

As you design your fitness program, keep these points in mind:

Consider your fitness goals. Having clear goals can help you gauge your progress and stay motivated. Create an exercise plan that incorporates a balanced routine. Get at least 150 minutes of moderate aerobic activity or seventy-five minutes of vigorous aerobic activity a week or a combination of moderate and vigorous exercise. To provide an even more significant health benefit and to assist with weight loss or maintaining weight loss, at least three hundred minutes a week is recommended.

Even small amounts of physical activity are helpful. Being active for short periods throughout the day can provide health benefits.

Do strength training exercises for all major muscle groups at least twice a week. Aim to do a single set of each exercise, using a heavy weight or resistance level to tire your muscles after twelve to fifteen repetitions.

Start low and progress slowly. If you're just beginning to exercise, start cautiously and go slow. If you have an injury or a medical condition, consult your doctor or an exercise therapist for help designing a fitness program that gradually improves your range of motion, strength, and endurance.

Build activity into your daily routine. Finding time to exercise can be a challenge. Schedule time to exercise as you would any other appointment to make it easier. Plan to watch your favorite show while walking on the treadmill, read while riding a stationary bike, or take a break to go on a walk at work. Small amounts of activity can add up: park your car at the far end of the parking lot and take the stairs at work instead of the elevator.

Plan to include different activities. Different activities (cross-training) can keep exercise boredom at bay. Plan to alternate activities that emphasize other body parts, such as walking, swimming, and strength training. Cross-training using low-impact forms of activity, such as biking or water exercise, also reduces your chances of injuring or overusing one specific muscle or joint.

Try high-interval intensity training. In high-interval intensity training, you perform short bursts of high-intensity activity separated by recovery periods of low-intensity exercise.

Allow time for recovery. Many people start exercising with frenzied zeal—working out too long or too intensely—and they give up when their muscles and joints become sore or injured. Plan time between sessions for your body to rest and recover.

Put it on paper. A written plan may encourage you to stay on track.

3. Get your equipment.

Start with running shoes. Be sure to pick shoes designed for the activity you have in mind. For example, running shoes are lighter in weight than more supportive cross-training shoes.

If you're planning to invest in exercise equipment, choose something practical, enjoyable, and easy to use. You may want to try out certain equipment at a fitness center before investing in your own equipment.

You might consider using fitness apps or other activity tracking devices, such as tracking your distance, recording burned calories, or monitoring your heart rate. (See Chapter 18 for more information.)

4. Get started.

Start slowly and build up gradually. Give yourself plenty of time to warm up and cool down with easy walking or gentle stretching. Then speed up to a pace you can continue for five to ten minutes without getting overly tired. As your stamina improves, gradually increase the amount of time you exercise. Work your way up to thirty to sixty minutes of exercise most days of the week.

Break things up. Any amount of activity is better than none. You don't have to do all your exercise at once, so you can weave in other activities throughout your day. Shorter but more frequent sessions have aerobic benefits, too. Exercising in short sessions a few times a day may fit your schedule better than a single thirty-minute session.

Be creative. Maybe your workout routine includes various activities, such as walking, bicycling, or rowing. But don't stop there. Take a weekend hike with your family or spend an evening ballroom dancing.

Listen to your body. You may be pushing yourself too hard. Take a break if you feel pain, shortness of breath, dizziness, or nausea.

Be flexible. If you're not feeling good, give yourself permission to take a day or two off.

5. Monitor your progress.

Retake your personal fitness assessment (that you did in Step 1) six weeks after you start your program and then again, every few months. You may notice that you need to increase your exercise time to continue improving. Or you may be pleasantly surprised to find that you're exercising just the right amount to meet your fitness goals.

If you lose motivation, set new goals or try a new activity. Starting an exercise program is an important decision. But it doesn't have to be an overwhelming one. By planning carefully and pacing yourself, you can establish a healthy habit that lasts a lifetime.

Warming up and Cooling Down

Every session of exercise should include a warm-up and cooldown. The warm-up period should *not* include static stretching. It should instead be a gradual increase in pace and intensity of the exercise. This allows the body to increase blood flow to the muscles and decreases the likelihood of a muscle or joint injury. The warm-up should last between five and ten minutes. The cool-down session should last a similar amount of time as the warm-up, with the pace gradually decreasing. Stretching exercises would be appropriate after physical activity.

In the Beginning: Baby Steps

When people start an exercise, they usually go overboard in their first few days. It puts lots of stress on the body and makes you feel burned out. And hence it seems like an unpleasant activity which you don't want to continue anymore. It's much better to start with fifteen to twenty minutes, slowly taking it and gradually increasing. Make the body adjust to activity so it feels good and progressively increases with your lifestyle.

Let's Check Back on Boyd!

Boyd embarked on an exercise program. He joined a local gym and arranged for a trainer to put him on a program for weight loss. Boyd was placed on a calorie-restricted diet. He lost about two pounds a week and reached his ten-week goal of a twenty-five-pound weight loss. Boyd was able to discontinue the use of his antihypertensive and cholesterol-lowering medications. His diabetes was under better control. And he reported an increase in the length of his penis! (Reducing the amount of abdominal fat/adipose tissue gave the appearance of his penis having an increased length.) He could see his toes for the first time when he took a shower! Boyd and his partner were much happier with their ability to engage in sexual intimacy.

Bottom Line

Exercise is beneficial for physical and mental health. It prevents and helps manage many diseases and can help us live a longer and healthier life.

References

1. Kemppainen, Susanna Maria, Lilian Fernandes Silva, Maria Anneli Lankinen, Ursula Schwab, and Markku Laakso. "Metabolite Signature of Physical Activity and the Risk of Type 2 Diabetes in 7271 Men." *Metabolites* 12, no. 1 (January 2022): 69.
2. Saint-Maurice, Pedro F., Barry I. Graubard, Richard P. Troiano, David Berrigan, Deborah A. Galuska, Janet E. Fulton, and Charles E. Matthews. "Estimated Number of Deaths Prevented through Increased Physical Activity Among US Adults." *JAMA Internal Medicine* 182, no. 3 (March 2022): 349–52. https://doi.org/10.1001/jamainternmed.2021.7755.
3. del Pozo Cruz, Borja, Matthew N. Ahmadi, I-Min Lee, and Emmanuel Stamatakis. "Prospective Associations of Daily Step Counts and Intensity with Cancer and Cardiovascular Disease Incidence and Mortality and All-Cause Mortality." *JAMA Internal Medicine* 182, no. 11 (2022): 1139–48.
4. Reimers, C. D., G. Knapp, and A. K. Reimers. "Does Physical Activity Increase Life Expectancy? A Review of the Literature." Special issue, *Journal of Aging Research* 2012. https://doi.org/10.1155/2012/243958.

CHAPTER 17

Meditation, Prayer, and Mindfulness

Roberto was a successful website developer who had many important and very demanding clients. He also had deadlines that he had to meet. As a result, his stress level was very high, and he found that he was gaining weight by not eating properly. He also suffered from insomnia and a mildly elevated blood pressure. A friend recommended he consider practicing meditation and mindfulness.

Women are overwhelmingly attracted to mindfulness; unfortunately, men aren't quite as interested. Why is this? The problem with mindfulness for men is that the practice asks them to become vulnerable. In prehistoric times, man's greatest responsibility was to protect his tribe. Our brains have been crafted over thousands and thousands of years to guard against vulnerability. Mindfulness also uses very feminine language like "warmth," "tenderness," and "gentleness."

In most cases, the physical threats that prehistoric men were guarding against are no longer the threats of modern-day life. Unfortunately, the brain hasn't caught up with the modern issues of threats, fears, and physical danger and thereby treats emotional vulnerability as a threat rather than an opportunity for men to truly reach their highest potential.

This chapter will define mindfulness, state its benefits, and provide the basic techniques for practicing this world view.

What is Mindfulness?

Mindfulness is the practice of purposely focusing your attention on the present moment—and accepting it without judgment.[1] Mindfulness focuses your attention on experiencing the present without judgment from the past or worries about the future. Mindfulness is now being examined scientifically as a critical element in stress reduction, overall happiness, and perhaps the control of serious medical conditions, including some cancers.

Historically, mindfulness has its roots in Buddhism, and it has been used for millennia to help control anxiety and depression. Most religions include some type of meditation technique that helps shift a follower's thoughts away from preoccupation with life's stresses and problems toward an appreciation of the current moment and a consideration of life's bigger picture.

Dr. Jon Kabat-Zinn, PhD, founder and former director of the Stress Reduction Clinic at the University of Massachusetts Medical Center, brought the practice of mindfulness and meditation into mainstream medicine. Dr. Kabat-Zinn demonstrated that practicing mindfulness can bring improvements in both physical and psychological symptoms and positive changes in health, attitudes, and behaviors.[2]

Mindfulness has the potential to enhance a man's well-being by increasing his capacity for supporting the many attitudes that contribute to a satisfying life. Being mindful makes it easier to savor the pleasures in life as they occur. The mindfulness practitioner becomes engaged in all of life's activities and thus creates a greater capacity to deal with adverse events. Many who practice mindfulness are less likely to worry about the future or have regrets over the past. They are also less preoccupied with concerns about success and self-esteem and can better form deep connections with others.

Mindfulness techniques help improve one's health. In addition to reducing anxiety, mindfulness can help relieve stress, treat heart disease, lower blood pressure, reduce chronic pain, improve sleep, and alleviate gastrointestinal difficulties. In recent years, mental health experts have recognized that mindfulness is an essential element in treating mental health problems, including depression, substance abuse, post-traumatic stress disorders, eating disorders, couples' conflicts, anxiety disorders, and obsessive-compulsive disorders.

How Does Mindfulness Work?

Mindfulness experts believe that mindfulness works, in part, by helping men to accept their experiences—including painful emotions—rather than denying those feelings and internalizing them, a response that can cause emotional havoc on a man and negative impacts for his family and friends.

Mindfulness results in changes in the brain and a decrease in the body's production of hormones that can suppress the immune system when increased. Research suggests that mindfulness leads to non-judgmental acceptance of any negative experience by replacing those negative thoughts with positive ones, resulting in psychological wellbeing.

Getting Started

There are several ways to practice mindfulness, but the goal of any mindfulness technique is to achieve a state of alertness and focused relaxation by deliberately paying attention to thoughts and sensations without judgment. This allows the mind to refocus on the present moment.

Most beginners of mindfulness focus on their breath. There are multiple breathing techniques, and you might consider either having an instructor or using one of the many apps or digital guides that provide directions for focusing on your breath (suggestions at the end of this chapter). You will also find the book *Breath*, by James Nestor, an excellent resource for learning a breathing method that is easy to learn and follow.[3]

Focused Breathing

For the most part, mindfulness breathing should take air in through the nostrils and exhale through the nostrils or the mouth. Inhaling through the nose is important because the nose allows filtering of the air, heating it, and moistening the air for easier absorption. One of the easiest methods for beginners is to use alternate nostril breathing. Basically, this is accomplished by placing your thumb over one nostril and inhaling through one nostril, pausing briefly, then placing the index finger on the opposite nostril and exhaling and then repeating this five to ten times.

Taking a full breath and expanding the chest and the abdomen is scientifically beneficial. When you breathe deeply, the air coming in through your nose fills your lungs, and you will notice that your chest expands, and your lower belly rises. Briefly, both lungs hold about 3,000 milliliters of air. When we are resting or doing sedentary activities, we inhale and exhale only 500 milliliters of air. As a result, most of our lungs' capacity isn't used, and stale air containing carbon dioxide accumulates in the lungs. By taking a deep breath and exhaling as much air as possible, we remove the carbon dioxide and increase the beneficial oxygen delivered to the bloodstream.

Square Breathing

One breathing technique used by the Navy SEALs is called "square breathing." You begin by inhaling to a count of four, holding your breath for a count of four, exhaling for a count of four, and holding for a count of four. Repeating this cycle of square breathing for five or six rounds will lead to relaxation, and it is especially effective before sleeping.

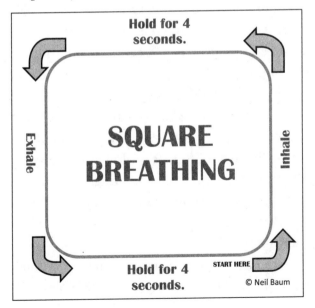

Figure 1: Square Breathing Technique

Petal Breathing

You can also achieve relaxation at any time of day using "petal breathing." This is accomplished by sitting, standing, or even lying down. Close your eyes and place your thumb on the four fingers of your hand like a closed flower. Take a deep breath and inhale through your nose, and at the same time, open your hand as if the petals of the flowers are opening to the sun. Then, exhale and allow the thumb to return to the tips of the fingers as if the petals of the flower are collapsing and coming together. Repeat this three times, and you will often find that you can achieve relaxation and remove the stressful situation you are experiencing.

Regardless of the mindfulness technique you select, you will usually find that your attention begins to wander after a few seconds. This happens to nearly every beginner who is learning mindfulness. It is challenging to stay focused, and our minds are easily distracted. Whenever you find your mind wandering and not focused on your breathing, slowly bring your attention back to your breath. This is not easy to accomplish, but it is a process that can be learned. It is almost like exercise and strength training. You can't expect to work out one time and have a muscular body or an abdominal six-pack. You must continue the exercise even when you are distracted. With repetition and focus on the exercises, muscular strength will occur. The same applies to mindfulness breathing and staying focused on your breath. With repetition and bringing the mind back to your breath, you can become proficient in breathing exercises.

Reciting a Mantra

Another method of staying focused is using a word, phrase, or mantra. Some mindfulness beginners find it easier to maintain their attention by slowly repeating a meaningful word, phrase, or mantra. A mantra is a single word or phrase that helps the mind from wandering. Repeating your word or mantra out loud can prevent wandering thoughts and allows focus on your breathing exercise. Any word or phrase can be used that is meaningful to you. Examples of a mantra include the Hindu Buddhist *Om,* the Judeo-Christian *Amen,* the Arabic *Salam,* and the Hebrew *Shema.*

Doing a Body Scan for Total Body Relaxation

The body scan technique helps you become attuned to your body and aware of your mind and body connection. A body scan can help you locate and release the tension in your body. This technique incorporates focusing on breathing and visualization.

Performing a body scan is quite simple. Begin by concentrating on one part of your body. Then try to focus on that muscle of that body part. Next, imagine those muscles as being warm and relaxed. This often results in any tension melting away.

As a guide, follow these steps, which are adapted from Dr. Herbert Benson and Aggie Casey's book *Mind Your Heart:*[4]

- Sit or lie down. Breathe deeply, allowing your stomach to rise as you inhale and allow your stomach to fall as you exhale. Breathe this way for two minutes before you start.
- Concentrate on your right big toe. Imagine the atoms in your toe and focus on the space between each atom. Imagine your toe feeling open, warm, and relaxed.

- Now shift your focus to each of the other toes on your right foot, visualizing them. Again, notice the sensations of your toes and envision them as open, warm, and relaxed.
- Slowly shift your focus to your foot, moving mentally from the ball of your foot to the arch, then the top of the foot.
- Now work your way up to your leg, turning your attention to your ankle, calf, knee, thigh, and hip. Take your time, slowly working through each area. For each body part, envision the atoms and the space between those atoms. Picture each muscle feeling open, warm, and relaxed.
- Allow your right leg to relax, sinking into the support of the floor.
- Now repeat these steps, focusing on your left foot and leg.
- Next, become aware of your back. Does it feel tight or tense? Pay attention to each vertebra and the space that surrounds it. Let each vertebra feel light and spacious. Slowly work your way up to your back, relaxing each muscle there.
- Gradually, move on to your abdomen and chest. Picture your organs and the space between them. Allow your belly to feel light and open.
- Become aware of your right thumb and then your remaining fingers. Envision each finger one by one, then slowly work your way through your hand and arm: relax your palm, wrist, forearm, elbow, upper arm, and shoulder.

- Feel your right arm relax and feel warm, spacious, and light.
- Do the same thing with your left hand and arm.
- Yawn. Shift your attention to the top and back of your head. Think about your neck and jaw. Allow each part of your face to relax, working through your jaw, eyes, and forehead.
- Let your body sink into your chair or bed. Does it feel light and relaxed? Focus on your breath. Imagine yourself breathing in calm and peace. Imagine any remaining tension being expelled from your body as you breathe out.
- If any part of your body is still tense, focus your breathing in that area, releasing tension from that spot as you exhale.
- Sit or lie quietly for a few minutes, noting how light and spacious your body feels. Then open your eyes slowly. Take a moment to stretch at the end of the body scan.

Roberto's Quest for Mindfulness

Roberto recognized that his work was becoming more stressful and that the stress and anxiety were spilling over to his relationship with others even when he wasn't at work. When he was at work, he thought about his home and his family. When he was with his family, he was distracted by thinking about his work. This was an unacceptable situation, and Roberto recognized that he had to change.

A friend recommended a meditation app, calm.com, and he used the petal breathing technique described above whenever he felt anxious at work. He found that just taking three to five deep breaths helped alleviate the anxiety occurring at work. Now he awakens each day and spends just a few minutes meditating before going to work. His doctor even reported that his blood pressure has returned to normal without medication.

Bottom Line

Stress and anxiety can affect you physically. Certainly, there is a role for mindfulness for men experiencing anxiety, an excess of negative thinking, or even depression. This is an option that is not expensive, has no side effects, and has a track record of being helpful. Breathe in, and begin!

References

1. Ludwig, David S., and Jon Kabat-Zinn. "Mindfulness in Medicine." JAMA 300, no. 11 (2008): 1350–52. https://doi.org/10.1001/jama.300.11.1350.
2. Kabat-Zinn, Jon. "Meditation Is Not What You Think." *Mindfulness* 12, no. 3 (March 2021): 784–87. https://doi.org/10.1007/s12671-020-01578-1.
3. Nestor, James. *Breath: The New Science of a Lost Art.* New York: Riverhead Books, 2020.
4. Change to Chill (website). "Head-to-Toe Relaxation." N.d. https://www.changetochill.org/wp-content/uploads/2018/08/Head-to-Toe-Relaxation-1.pdf.

CHAPTER 18

Home Testing and Wearables

A urologist has seen Ivan for lethargy and a decrease in his sex drive or libido. Ivan also has adult-onset diabetes mellitus and has to draw a drop of blood from his finger multiple times a day. Ivan had lab tests that confirmed low testosterone levels. He was treated with injections of testosterone, administered by his nurse wife every two weeks. Ivan had a normal digital rectal exam and a normal PSA level. He was advised to obtain a blood count to check for an increase in red cells or polycythemia, a PSA test, and a follow-up of his testosterone level. Ivan claimed that it was inconvenient for him to go to the lab to discuss the results with his urologist.

At-Home Medical Tests

We have learned from the global pandemic that it is possible to provide medical care using telemedicine (see Chapter 19). Medical care has moved from the hospital to the doctor's office and now from the office to the patient's home. It is not necessary to be face-to-face with the patient in many situations. Currently, lab tests can be conducted in the comfort of the patient's home. This chapter will review the concept of home testing, what tests are applicable for home testing, and the advantages and disadvantages of home testing.

At-home tests for detecting COVID-19 have received an abundance of publicity. In December 2021, with the Omicron variant of the coronavirus spreading quickly, the government agreed to have COVID-19 tests delivered to Americans at their home—at no cost.

The COVID-19 tests are only one of the many available home medical tests. Users generally collect a "sample"—typically blood, urine, saliva, or mucus—and get immediate results or send it to a lab the test manufacturer designates. The most helpful home medical tests may be those that help people with chronic conditions such as diabetes, congestive heart failure, and high cholesterol levels.

In the past, this home testing has been used to diagnose an illness or monitor glucose levels in diabetic patients. However, thousands of new tests have become available in recent years, showing up on store shelves and on the internet—many from companies like Everlywell (everlywell.com), LetsGetChecked (letsgetchecked.com), and myLab Box (mylabbox.com). Some of these home tests are straightforward, such as those for COVID-19, but others have questionable metrics, like "cell aging."

Unfortunately, most of these products aren't covered by insurance. The cost can range from less than $10 for strips to check urine for bacteria to $1,000-plus for genetic testing. Also, the quality of some of these home tests is unreliable, and some may have confusing results. When the results are misleading, additional follow-up tests, a delay in care, and otherwise unnecessary treatments may be recommended.

The Food and Drug Administration has authorized the marketing of more than one hundred categories of home medical tests. (To find the list of authorized tests, go to fda.gov, search for "in vitro diagnostics," then click on "Home Use Tests"). Some of these home-testing kits have been reviewed by the FDA to certify that they can measure the manufacturer's claims accurately and reliably.

The FDA has also approved a handful of products for diagnosing issues such as urinary tract infections. And the Centers for Disease Control and Prevention recommends that people consider using a rapid self-test for COVID-19 before joining any gatherings in public or away from home.

If you have mild symptoms, your doctor may be able to use home testing followed by a telemedicine or phone discussion to review the results and recommend treatment. Also helpful is an FDA-approved home test for HIV, which is key for people without access to a health care provider or have concerns about privacy. And with your doctor's okay, you can use a home fecal test to screen for colon cancer or a small blood sample to screen for hepatitis C.

Home-Testing for Men

Dozens of home-testing products are specified for men. The most common at-home testing kits for men are those used to detect testosterone level, PSA level, and colon cancer.

The PSA test is a screening test for detecting prostate cancer or following the man after treatment for prostate cancer. The home test requires a finger-tip puncture to obtain a few drops of blood. Men should not ejaculate before the test since the semen released can elevate PSA levels and skew the results. We suggest a man abstain from sexual activity—either masturbation or sexual intimacy with a partner—for forty-eight hours before the test. Most at-home PSA kits recommend collecting blood samples first thing in the morning, but no fasting is required.

Testosterone home testing kits are widely available from several companies, such as LetsGetChecked (letsgetchecked. com), Everlywell (everlywell.com), and Progene (progene.com). They use your blood or saliva to test your testosterone levels. Testosterone levels should be taken in the morning when the testosterone level in the blood is the highest. There is no limitation on food before the test. Before the test, men don't have to abstain from intercourse or masturbation.

Screening for colon cancer can be done with a home-testing kit that looks for small amounts of blood in the stool or feces. Additional studies are recommended if blood is present in the stool, even in small quantities that cannot be seen with the naked eye. Some foods or drugs can affect the results of this test. This includes nonsteroidial anti-inflammatory drugs (NSAIDs), such as ibuprofen (Advil), naproxen (Aleve), or aspirin, for seven days before testing. (They can cause bleeding, leading to a false-positive result.) Vitamin C of more than 250 mg a day from either supplements or citrus fruits and juices for three to seven days before testing can affect the chemicals in the test and make a false negative result. We also recommend avoiding red meats (beef, lamb, or liver) for three days before testing. (Components of blood in the meat may cause a positive test result.)

Drawbacks of Home-Testing

Like everything else in medicine, nothing is perfect or 100 percent reliable. Tests that the FDA hasn't okayed can have several drawbacks.

The FDA doesn't review "wellness" tests. These are used to measure hormone levels, food sensitivities, general heart health, blood levels of vitamins, stress, and cell aging. These tests are not intended to diagnose specific conditions. In addition, the agency typically does not check "laboratory-developed tests" (LDTs), which are developed and used by a single lab. But the FDA has been paying attention to LDTs. A 2018 statement identified potential problems, such as claims not supported by evidence, erroneous results, and faked data. Some home testing companies, like Everlywell (everlywell.com), say that their tests are laboratory-developed tests or LDTs. If the FDA has reviewed a test, that fact is likely to be included in the company's marketing materials.

Certain top-selling home tests aim to identify food sensitivities by checking a user's blood sample for IgG, an immune system antibody. However, the leaders in allergy and asthma have not confirmed evidence linking IgG levels to food sensitivities and allergies. The FDA has also warned about home genetic tests that manufacturers claim will predict how your body would respond to antidepressants, heart drugs, and other medications.

Home tests for male hormones (testosterone) are plentiful. But knowing your hormone levels doesn't necessarily pinpoint why you feel depressed or exhausted. Numerous health issues, including anemia, depression, infections, and sleep apnea, can cause fatigue, depression, and blunting of the sex drive.

Home testing for genetic screening of the risk of Alzheimer's disease, cancer, and other serious conditions is particularly concerning. These tests can't tell you whether you will develop an illness or give advice other than to follow existing health guidelines.

Some manufacturers make health care professionals available to recommend tests, counsel users, and even prescribe medications. But they can have a vested interest in the testing company they represent. These online consults also lack information about you and your medical history. And factors such as your age and the medications you take can affect a home test's results. We suggest consulting your doctor before using a home medical test. A doctor familiar with your medical background is likely to have a greater understanding of how to identify medical problems than a single home medical test.

A few recommendations

Start by asking your doctor if home testing is the best way to get medical information. In most cases, your doctor's orders are typically covered by insurance; most of those you can request online are not. There might be an alternative approach that quickly gets to the bottom of your medical problem.

Check the label or description for home testing kits you send out to make sure the lab is "CAP accredited" or "CLIA certified." This means that the test meets quality standards and that the lab undergoes regular inspections.

Be sure to check storage directions and the expiration date. Some tests are sensitive to temperature and humidity. Of course, don't use testing kits that have expired. Also, note that time of day, food and drink consumed, and supplements you take can affect results. Many home testing kits include frequently asked questions or FAQs to guide you.

Please note that no test is 100 percent accurate. For example, COVID-19 tests that provide immediate results are generally less sensitive than those you send to a lab, and home tests for a urinary tract infection can't detect less common types of bacteria.

At-Home Genetic Testing

In 2003, excitement about the future of genetic science and its potential to dramatically advance the course of disease prevention and treatment was running high. The National Institutes of Health (NIH) announced that researchers had completed the first whole sequence of the human genome. The hope of identifying and locating every gene would lead to an understanding of the inherited risks for cancer, improve care for conditions like diabetes, and help scientists develop targeted therapies for disorders such as Alzheimer's.

Genomics has led to critical progress in medical science almost two decades later, particularly in identifying an individual's genetic predispositions to diseases such as certain forms of cancer. This knowledge has led to many consumer products in the form of at-home genetic tests.

Anyone can purchase a home genetic test kit for a relatively modest fee. Brands such as 23andMe and Ancestry are household names. Users spit into a tube or swab the inside of their mouth to obtain genetic material, return the sample by mail, and receive an analysis a few weeks later. Claims from the manufacturers of these home-testing products are important to check out.

About one in five Americans has taken a home genetic test.[1] Some manufacturers say that their tests can help assess whether you are more likely to develop a disease such as Alzheimer's.

The disease risk results that you get from a home-testing kit might be less complete than that of a test you get as part of your regular medical care. In contrast to at-home testing, your health care providers will generally work with you to determine whether you need additional tests based on your health history. They'll also connect you with genetic counselors, who can help ensure that you're tested for potentially relevant genetic variants.

Consider also that home testing might not include a wide range of important gene variations. Take, for instance, 23andMe's (23andme.com) test for BRCA1 and BRCA2 genes, which have been linked to a higher risk of prostate cancer.[2] If a man tests positive for this genetic variant, close follow-up with a urologist is indicated.

A negative result may provide a false sense of relief because it doesn't look at all the variants. Also, home testing manufacturers may offer genetic counseling. However, the company's counselor will be less familiar with your medical history than your doctor or a genetic counselor recommended by your doctor.

Several home-testing companies currently offer personalized diet recommendations based on your genetics. Some draw their advice from a specific genetic variation that may be linked to a better response to a specific diet or exercise program, but research about gene variants and diet is observational and hasn't been adequately replicated in other studies.

Home genetic tests can provide certain insights, but it's important to be aware of their limitations. Keep in mind that your genetics are only one element of your overall well-being. Your environment, health care access, and behavior are also important.

Wearable Technology

Imagine if you and/or your healthcare provider could get potentially life-saving information in real time. Today, this technology is possible, and you are probably already using it to monitor health and fitness parameters.

Wearable technology, also known simply as wearables, are electronic devices that are worn on the body. These tools utilize microchips to gather, interpret, and transmit information (data) to the wearer or to a health care provider. Examples of wearables include fitness trackers and smartwatches, but they can also be sensors that monitor your heart rate, electrocardiogram, oxygen levels, glucose levels, or a variety of other health or fitness measurements.

The market for wearables is booming. In 2021, global consumers were expected to spend $81.5 billion dollars for these devices, an increase of 18 percent from the $69 billion spent in 2020 (Gartner.com). The popularity of these products also continues to grow. In 2022, more than 80 million people worldwide were expected to use a wearable (Insiderintelligence.com).

History of Wearables

Wearable technology has been available for much longer than you probably realize. The first of these devices was probably a watch that also functioned as a calendar and calculator. That innovation was produced by Hewlett-Packard in the late 1970s. Also, you might remember the Sony Walkman that was popular during the 1980s. The Walkman allowed users to listen to music on the go. It also was the early prototype for modern digital music players. Smartwatches made their debut in the 1990s and included the Timex Datalink and the Linux Smartwatch. Although not the first MP3 player, music fans remember the release of the first Apple iPod in 2001. Other devices that have led to the modern use of wearables include the Fitbit (introduced in 2009), and the Apple Watch (introduced in 2015.)

Types of Wearables and Their Applications

The variety of smart wearables available and the applications of these devices has continued to grow. As technology improves, these sensors can be worn on virtually any part of the human body. Some of the products and uses are listed below.

Fitness Trackers

Fitness trackers (also called activity trackers) were developed to assist consumers to increase movement and/or exercise. These devices are worn on the wrist and—depending on the device—can count steps, calories burned, keep track of exercise time and type, and monitor heart rate and blood pressure. Some fitness trackers also have sleep monitoring and global positioning system (GPS) capabilities. For the past ten to fifteen years, these devices have helped people increase activity and sometimes lose weight.

Smartwatches

Smartwatches evolved from fitness trackers; they usually incorporate the health benefits of a fitness tracker into a product that has additional applications. These watches are basically wearable smartphones, and they incorporate mobile apps as well as digital music and Bluetooth connectivity. Health applications of these watches may include those offered by fitness trackers plus additional features such as the ability to monitor blood oxygen levels, electrocardiograms, body temperature, respiratory rate, and blood sugar.

Continuous Glucose Monitors (CGMs)

Recently, a colleague enjoying a destination spa noticed that many of her fellow guests were sporting white discs on their upper arms. After an inquiry, she discovered that they were wearing continuous glucose monitors, or CGMs.

The first CGM was approved by the FDA in 1999, and since that time these devices have improved. The current products allow for monitoring of blood glucose levels twenty-four hours a day and use a sensor placed under the skin. A transmitter sends the readings to a cellphone app, wearable receiver, or insulin pump. Traditionally, CGMs were developed to assist type 1 and type 2 diabetic patients in monitoring trends in glucose levels. In the past, diabetic patients performed multiple finger sticks every day to access glucose levels. In addition to being painful, finger sticks only show glucose levels at the exact time that they are done. Glucose levels change with food, activity, illness, and stress.

The CGM helps diabetic patients and their health care providers with dosing of insulin and other medications used in the treatment of diabetes. In addition to fewer finger sticks, other benefits of CGMs in diabetic patients include alerts for hypoglycemia and a better quality of life for patients. Be aware that some finger sticks are still needed to calibrate the device.

These wearables may also have health benefits in non-diabetic individuals, including the earlier diagnosis of diabetes or prediabetes in people with risk factors. In the past, fasting blood glucose is a common testing modality used by healthcare providers, but because it only measures the glucose level at one moment in time, elevated levels at other times during the day might be missed. By continuously monitoring the levels, it may be easier to identify times (such as after meals) that blood glucose levels are elevated.

A second application of CGMs for non-diabetic users is to help guide dietary choices. For example, the CGM shows that the blood glucose rises after eating a sugary food such as a donut or piece of cake and thus helps the users make healthier food choices. Use of CGMs can also help encourage individuals to initiate or continue an exercise plan. Exercise lowers blood glucose and can help decrease risk factors for obesity, heart disease, and other disorders such as hypertension which are linked to a sedentary lifestyle. Users have the data available in real time to motivate them to exercise.

Sleep Trackers

Adequate sleep is an instrumental part of a healthy lifestyle. Adults should get seven to nine hours of sleep every night. Unfortunately, many people don't get enough sleep. According to the CDC, about one-third of adults in the US sleep fewer than seven hours per night.

Why is sleep so important? Lack of sleep has been associated with adverse effects on health and increases the risk of disorders such as obesity, heart disease, cancer, and diabetes. In addition to physical health problems, inadequate sleep is also related to mental health disorders such as depression. (cdc.gov)

Many of us know that we aren't getting the sleep that we need, but don't know why. Sleep trackers can't help us sleep longer or better, but may provide valuable information about sleep patterns that can assist in improving our sleep habits. Wearable sleep trackers are available in several forms including wristband, ring, chest strap, earbuds, and the aforementioned smartwatch. These devices can monitor sleep duration, quality, and phases. Some sleep trackers can also detect heart rate, respiratory rate, and even oxygen saturation. Some devices can even monitor environmental factors such as room temperature or amount of light in the bedroom. Users can use the data to determine how best to improve the quality and quantity of sleep. (See Chapter 20 on the importance of sleep.)

Novel Wearables: Current and Future

Smart Patches

Smart patches are wearables that are attached to the skin with an adhesive. These devices have multiple applications for health, including the following issues: vital sign (blood pressure, heart rate, body temperature) monitoring, cardiovascular (ECG) monitoring, diabetes management, medication delivery, stress management, fertility monitoring, wound care, and sleep tracking. The smart patch market was valued at more than $8 billion in 2020 and is expected to reach $15 billion by 2030 (biospace.com).

Smart Clothing

Your clothing may be getting smarter. Smart clothes are also known as e-textiles and are emerging as an innovative way to monitor your health and fitness. Sensors can be embedded into or attached to a garment, or the fabric itself may function as a sensor. Examples of smart clothing currently on the market include yoga pants, shirts, a compression sleeve, socks, swimsuits, sleepwear, and shoes.

Health applications of e-textiles include vital signs, diabetes, cardiovascular, and sleep monitoring as well as ultra-violet (UV) light exposure. Fitness applications encompass parameters such as calories burned, steps, speed, distance, heart rate, and workout intensity. Some clothing can reduce heat (shirt) or encourage proper technique (yoga pants). Stay tuned for future products!

Smart Earphones

Smart earphones are also known as "hearables." These devices can be used to enhance hearing for those with partial hearing loss. However, it may surprise you that hearables can also help with health and fitness. The outer portion of our ear is close to blood vessels and nerves, which can be used to assess biometric parameters such as heart rate, respiration rate, glucose levels, oxygen levels, blood pressure, and brain activity. In addition, hearables may be programmed to deliver sounds, music, or programming that might assist in sleep or with the reduction of mental health symptoms.

Advantages and Disadvantages of Wearables

In the health and fitness arena, wearables have several benefits:

- Ease of use
- Real-time monitoring that can be seen by users and healthcare providers
- Improving patient outcomes
- Encouraging and reinforcing healthy decision-making
- Shared decision-making between users and health care providers

There are also some concerns about widespread use and availability of these devices:

- Privacy and data security concerns

- Challenges with the technology, especially among elderly users
- Battery life of the devices
- Cost
- Accessibility
- Integration into electronic health record (EHR)

Update on Ivan's Health Care

In a follow-up telemedicine visit, Ivan's urologist recommended a home-testing company to determine Ivan's PSA and testosterone levels. Ivan did the tests in his home, mailed back the specimens to the company, and received the results in a few days. Ivan made an appointment for a telemedicine follow-up with his urologist, thus avoiding multiple trips to the doctor. Since Ivan's doctor ordered the tests, the insurance company paid for the testing. Ivan also met with his internist who was monitoring his diabetes mellitus. This doctor suggested a continuous glucose monitoring device. Now Ivan was able to control his insulin levels without the necessity of drawing blood from his finger. Ivan and his doctors embraced these new technological advances that ultimately improved his quality of life.

Bottom Line

At-home medical tests can provide helpful information, but they are not a replacement for guidance and treatments by a health care provider. After at-home testing, you should follow up with your provider, regardless of the result.

Technology has continued to transform our lives. Smart wearables are used by millions worldwide, and the market continues to grow. The potential uses for wearables for fitness and health are limitless. These devices are helpful for users to guide self-care and to modify risk factors that lead to an unhealthy lifestyle. Wearables also have the potential to revolutionize the way we prevent, monitor, and treat conditions like diabetes, arrhythmias, asthma, COPD, and high blood pressure.

References

1. Regalado, Antonio. "More Than 26 Million People Have Taken an At-Home Ancestry Test." *MIT Technology Review,* February 11, 2019. https://www.technologyreview.com/2019/02/11/103446/more-than-26-million-people-have-taken-an-at-home-ancestry-test/.
2. Nicolosi, Piper, Elisa Ledet E, Shan Yang, Scott Michalski, Brandy Freschi, Erin O'Leary, Edward D. Esplin, Robert L. Nussbaum, and Oliver Sartor. "Prevalence of Germline Variants in Prostate Cancer and Implications for Current Genetic Testing Guidelines." *JAMA Oncology* 5, no. 4 (April 2019): 523–28. https://doi.org/10.1001/jamaoncol.2018.6760.

CHAPTER 19

Telemedicine

Alonso is a CPA with a history of erectile dysfunction and low testosterone levels. Taking time from his busy accounting practice has been difficult. He has friends who communicate with their physicians through telemedicine. Alonso is interested in this concept and does not know how to find a doctor willing to provide virtual care.

The traditional doctor-patient relationship requires that the patient makes an appointment to see a doctor. The patient might have to wait weeks or months for that appointment, finally traveling to the doctor's office and waiting for a lengthy time in both the reception area and the exam room. The doctor enters the exam room, obtains a medical history, and examines the patient. The doctor writes a prescription, and the patient then travels to the pharmacy and waits for the prescription to be filled. This method of delivering health care can take several hours of a patient's time. This scenario is all too familiar to most patients seeking health care, and the process is inefficient for the patient and the doctor.

Compare that method of receiving care to today. A patient can call the physician's office and request a video conference or telemedicine visit. A virtual meeting is arranged on the day the patient calls or within a few days. The patient doesn't leave home or office but logs on to the doctor's computer at the designated time.

A video visit may last fifteen to thirty minutes, depending on the nature of the problem. The doctor electronically sends the prescription to the pharmacy. The patient can pick up the medicine or can request that the medication be delivered to the patient. The doctor may request that the patient report to a lab or imaging center for additional tests. This process, not including any lab or imaging studies, takes less than one hour. It is efficient for both patient and doctor.

The coronavirus pandemic has forever changed the way medicine is practiced. We have learned that receiving care for many medical conditions and problems is possible without being face-to-face with a doctor.

Most doctors and patients have practiced a form of telemedicine for decades; for example, when a patient contacts a physician via telephone to request medication refills. In these cases, physicians can either call the pharmacy to refill the medication or suggest that the patient make an office appointment to receive a new prescription.

Unfortunately, in this system of phone calls for advice or prescription refills, the physician receives no compensation. Seldom is there a record in the patient's chart of these phone calls from patients to their physician. Nevertheless, when telephone medicine is informally practiced, the physicians are legally responsible for their actions and advice. This does not represent good medical practice.

Telemedicine can enhance the communication and needs of patients who want to receive care through this convenient technology. Telemedicine saves patients the time and effort of coming to the office, and the telemedicine visit documents the interaction in the patient's record, making for better accountability. Many patients do not require an in-person visit to receive their medical care.

Conditions amenable to telemedicine include:

- Monitoring blood pressure
- Cholesterol testing
- Managing the well-controlled diabetic patient
- Uncomplicated urinary tract infections
- Mild upper respiratory tract infections
- Mild gastrointestinal symptoms
- Monitoring PSA levels in men who have had a negative digital rectal exam
- Monitoring testosterone levels
- Managing patient on medication for an enlarged prostate gland
- Some postoperative patients

Preparing for a Telemedicine Visit

You will probably be asked to sign a consent giving the doctor permission to treat you using video technology. (A sample video consent is shown in Figure 1.) You want to have your telemedicine visit in a quiet, private, distraction-free environment. For the best experience, think of your telemedicine appointment as an office visit at your physician's office.

Suggestions to prepare for your telemedicine appointment:
Check your equipment

- For smartphones, check your camera and microphone. You should face the camera or the lens on your computer. It's commonly referred to as Selfie Mode. Check that your battery is sufficiently charged on your smartphone.
- For computers, verify your speakers, microphone, and camera are functioning.
- Verify that your volume is turned up so you can hear the physician.
- Close all other applications/apps. This will improve the video quality and reduce battery consumption.

Use your smartphone on a Wi-Fi network to improve the video and audio quality. If you are unfamiliar with your equipment, ask a family member or friend to assist you

Prepare Your Space

- Choose a quiet, private, well-lit room for the visit. You'll be reviewing personal medical information, so make sure you have privacy for the visit.
- Avoid having your back to a bright window or other distracting backgrounds, and avoid backlighting, which makes your image too dark and indistinct.
- If available, use a mobile phone stand on a solid surface. It's best to set the mobile phone stand at eye level. Avoid moving the phone or camera during the visit unless the physician needs you to.

Reduce Distractions

- Set your phone to silent and, if possible, turn off other application notifications.
- Turn off other devices that may cause distractions during the visit, such as TVs, radios, or household appliances.
- If you have family members or guests in the house with you, ask them in advance to limit conversation during your visit.
- If you have pets, you might consider moving them to another location during the video visit.
- Consider using a headset during the visit so you and your physician can easily hear one another.
- We suggest you do not conduct your video visit in your vehicle, especially while driving! A car may not be appropriate for your type of visit, and your physician may need to ask you to change position during the visit.
- Avoid walking and moving around the room. While we have grown accustomed to various video calls with friends and family, a visit with your physician should be held in a quiet, private, well-lit room while sitting.
- We also suggest you do not have your visit at work in your cubicle, break room, or another common area. We also do not want you to get in trouble with your employer! Your employer may support your telemedicine visit and provide a quiet, private room for you to have it. Check with your employer *before* you schedule a video visit.

- Never connect to your appointment in a restaurant, gym, park, or other public location.
- Avoid multitasking. That means not responding to text messages, looking at emails, answering texts, emails, or your phone during your visit.

So, How's Alonso These Days?

Alonso called his doctor, who was treating him for erectile dysfunction and low testosterone levels. He inquired if the doctor would be agreeable to providing care using telemedicine. The doctor indicated this was possible after Alonso signed an informed consent. The doctor and Alonso found several times that were convenient for both. Alonso would have his lab work drawn before each telemedicine visit, and the doctor reviewed the results and answered any questions that Alonso had. Alonso agreed to come in for an office visit once a year. This arrangement was acceptable to Alonso; it allowed him to have medical care without taking several hours to achieve his desired results.

Bottom Line

Telemedicine is the future of health care in America. Patients want ready access to their health care providers without having to devote hours to a medical encounter that could be completed in minutes via telemedicine. This is a win-win solution for the patient and the physician. The benefits of telehealth include convenience, access to care, better patient outcomes, and a more efficient healthcare system.

It is not *if* telemedicine will become the most used communication method with the healthcare system, but *when* this will occur. This transition from eyeball-to-eyeball with the doctor compared to the eyeball-to-video screen afforded by telemedicine is already on the horizon and is growing in frequency. You may have already experienced this technological convenience!

Figure 1: Sample Telemedicine Consent Form

By signing this form, I understand and agree with the following: Telehealth/Telemedicine involves the use of electronic communications to enable health care providers at different locations to share individual patient medical information for the purpose of improving patient care. Providers may include primary care practitioners, specialists and/or subspecialists, nurse practitioners, registered nurses, medical assistants, and other health care providers who are part of my clinical care team. In addition to myself and the members of my clinical care team, my family members, caregivers, or other legal representatives or guardians may join and participate on the telehealth/telemedicine service, and I agree to share my personal information with such family members, caregivers, legal representatives, or guardians. The information may be used for diagnosis, therapy, follow-up and/or education.

Telehealth/Telemedicine requires transmission, via internet or tele-communication device, of health information, which may include:

- Progress reports, assessments, or other intervention-related documents
- Bio-physiological data transmitted electronically
- Videos, pictures, text messages, audio, and any digital form of data

The laws that protect the privacy and confidentiality of health and care information also apply to telehealth/telemedicine. Information obtained during telehealth/telemedicine that identifies me will not be given to anyone without my consent except for the purposes of treatment, education, billing, and healthcare operations. By agreeing to use the telehealth/telemedicine services, I am consenting to [NAME] sharing of my protected health information with certain third parties as more fully described in [NAME] Privacy Policy. I understand, agree, and expressly consent to [NAME] obtaining, using, storing, and disseminating to necessary third parties, information about me, including my image, as necessary to provide the telehealth/telemedicine services.

Telehealth/telemedicine sessions may not always be possible. Disruptions of signals or problems with the internet's infrastructure may cause broadcast and reception problems (e.g., poor picture or sound quality, dropped connections, audio interference) that prevent effective interaction between consulting clinician(s), participant, patient, or care team.

I hereby release and hold harmless [NAME] and all members of my care team from any loss of data or information due to technical failures associated with the telehealth/telemedicine service.

I have the right to withhold or withdraw consent to the use of telehealth/telemedicine services at any time and revert to traditional in-person clinic services. I understand that if I withdraw my consent for telehealth/telemedicine, it will not affect any future services or care benefits to which I am entitled.

I hereby consent to the use of telehealth/telemedicine in the provision of care and the above terms and conditions.

I have had all my questions answered. I understand that this informed consent will become a part of my medical record.

Signature of Patient _____ Date and Time _____

Printed Name of Patient _____

CHAPTER 20

The Importance of Sleep

Demarcus was suddenly awakened by the sound of a loud horn. He had crossed the double yellow line on the way home from a late evening at work. Fortunately, he was able to avoid a head-on collision. After sharing this close encounter with his wife later that evening, he decided to reexamine his life. In his quest to be more "productive" by putting in extra hours, he had become less effective at both work and at life in general.

Sleep deprivation is a growing yet poorly recognized problem of the twenty-first century. Not only does lack of sleep make you feel lousy, but it also poses a significant threat to your health and longevity. As many as one-third of adults report sleeping six or less hours per night (the definition of chronic sleep deprivation). More than 20 percent of adults suffer from a clinical sleep disorder such as sleep apnea. Regardless of the cause, lack of sleep can lead to medical conditions that further hinder our ability to sleep.

Many men feel that sleep involves "doing nothing." On the contrary, this is the time during which both the mind and body recharge themselves. For example, during this downtime, our body is hard at work repairing the damage that our cells endure throughout the day. Of course, without adequate sleep, our brain cannot function properly.

How Does Sleep Work?

Sleep is controlled by two factors—our need for sleep and our internal clock. Of course, the need for sleep is triggered by becoming tired. However, if we are obtaining enough sleep, this feeling should only occur close to bedtime. Our internal clock—the "circadian rhythm"—times our sleep to occur at night through a complex release of specialized hormones. Circadian rhythm is also synchronized to our environment, most notably lack of light.

We also tend to sleep in 90- to 120-minute cycles, each consisting of four stages of sleep. The fourth stage is called "REM" sleep, named for the rapid eye movement that occurs during this portion of sleep. Both dreaming and body paralysis occur during this stage, thereby allowing a high level of cognitive processing and memory consolidation. "Slow-wave" sleep is the stage from which we gain a refreshed feeling in the morning. Interruption of a sleep cycle will often lead to a feeling of tiredness. As such, even losing one hour of sleep can lead to significant sleep deprivation.

So, how much sleep do we need? Although everyone is different, most studies show that seven hours per night is the minimum requirement, with more hours needed for children. For those who require more sleep, typical indicators include difficulty waking up in the morning and feeling tired during the day.

Why Is Sleep Deprivation Dangerous?

The three pillars of great health are a nutritious diet, consistent exercise, and regular sleep. First and foremost, lack of sleep makes us more accident-prone. For instance, inadequate sleep can impair your ability to drive the same way as drinking too much alcohol. We also become more susceptible to falls and sports injuries. Although not as noticeable, sleep deprivation can lead to a whole host of medical conditions.

The Effects of Cortisol

Cortisol is an essential hormone that helps us combat stressful events. Some of its effects include increase in blood sugar and heart rate. When we are sleep-deprived, this level can remain elevated for prolonged periods of time, resulting in a myriad of undesirable effects.

When cortisol elevates our blood sugar levels, our body will produce insulin—a hormone that is secreted by our pancreas to bring this sugar into our cells. We store additional sugar as glycogen located in our liver and muscles. When we eat carbohydrates, our glycogen stores are replenished. Excess carbohydrates will be stored as fat. To make matters worse, our body will usually choose sugar for energy rather than excess fat. This vicious cycle will cause our cells to become tolerant to insulin, like the way we can become tolerant to pain meds with chronic use. Insulin resistance will then lead to the most harmful type of fat, "belly fat," also referred to as "visceral fat" because it surrounds our internal organs.

Independent of weight gain, cortisol causes a direct negative effect on our blood vessels, thereby stressing our heart and increasing our risk of high blood pressure, heart attack, and stroke. Although cortisol is an anti-inflammatory, poor sleep triggers chronic inflammation that can lead to plaque build-up in our arteries. Consistently sleeping less than six hours per night can increase your risk of heart attack by as much as 20 percent.

Mood Alteration

Poor sleeping habits can lead to a tenfold increase in the risk for developing clinical depression. In addition, men with depression commonly have difficulty sleeping. Furthermore, antidepressant medications can suppress REM sleep. Like depression, anxiety can both result from and contribute to sleep disturbances.

Immune System Dysfunction

Sleeping has a profound effect on the integrity of our immune system and its ability to combat illness. The immune system's production of specialized cells and chemicals decreases dramatically in the absence of adequate sleep. It is therefore no surprise that sleep deprivation increases our risk of viral illnesses and cancers.

Benefits of a Good Night's Sleep

In addition to reducing our risk of developing serious medical conditions, a good night's sleep can improve our overall quality of life. Sometimes it's easier to focus on the immediate gratification benefits rather than the hidden ones.

Mental Resilience

In addition to reducing stress, adequate sleep allows us to cope with the demanding tasks that we face throughout the day.

While we sleep, we can sort the complex information that our brain received during the previous day, thereby improving our concentration, creativity, and learning ability the following day.

Improved Heart and Brain Health

When we lie down in bed each night, we are not thinking about how rest reduces our risk of heart attack and stroke. However, we all know that a good night's sleep will allow us to function better the next day. So, what makes us feel sleepy? When our body produces chemicals such as melatonin and adenosine, we begin to feel tired as bedtime approaches. However, during the day, that tired feeling is partly due to a decrease in cognitive and cardiovascular function.

Being in a Better Mood

Having a pleasant disposition is essential for maintaining successful relationships at home and at work. Adequate sleep will reduce irritability, frustration, and anger. Of course, mood swings can also affect sleep quality, thereby creating a vicious cycle.

Increased Productivity

Financial investors often use the phrase, "Earn money while you sleep." This term usually refers to passive income derived from investments in an appreciating asset or an automated process. Conversely, burning the midnight oil leads to diminishing returns in productivity. Investing in a good night's sleep will pay a multiple of dividends in productivity throughout our waking hours by improving our cognitive function and memory.

Improved Exercise Tolerance and Athleticism

Ask any professional athlete, and they will tell you that sleep is an essential part of their regimen. Sleep also reduces the risk of sports-related injuries. More importantly, sleep deprivation will often cause us to neglect our exercise routine.

Easier Weight Control

Hormonal imbalances associated with sleep deprivation can lead to weight gain. Some of these imbalances cause a decrease in our production of chemicals that naturally suppress appetite. We also tend to make poor dietary choices when we are tired.

Healthy Tips for Healthy Sleep

The Right Environment

Perhaps cavemen slept better than the modern man. After all, the ideal sleep environment is cold, dark, and quiet. The bedroom should also be a dedicated space for sleep and intimacy. Work and hobby-related items can disturb our natural wind-down process.

Ritual

As mentioned earlier, our sleep cycle depends on a circadian rhythm. Having a consistent bedtime routine and time will reinforce our internal clock's signal for entering the first stage of sleep and keeping us asleep. Bedtime rituals can include a cup of herbal tea, a warm bath, or some light reading. Avoid electronic devices, TV, and eating for the last couple of hours before going to bed.

Exercise

Consistent exercise will lead to a better night's sleep. However, exercise should be avoided in the few hours leading up to bedtime. The good news is that it doesn't take much exercise to realize the benefit. As little as a twenty-minute walk at a moderate pace can make a noticeable improvement.

Weight Control and Other Dietary Concerns

Moderating the amount of sugar in our diet can temper some of the hormones that lead to sleep disturbances. Reducing sugar intake can decrease excess belly fat that is often associated with the most common sleep disorder, sleep apnea. Minimizing consumption of alcohol and caffeine will also lead to a better night's sleep. The timing of when to discontinue caffeine differs with the individual. For some, even caffeine in the middle of the day can disrupt our sleep patterns. Also, the need for caffeine can be a signal of sleep deprivation.

Apps, Wearables, and Other Devices

Sometimes having a pad of paper and a pen on the nightstand can help us sleep better. When we write something down, it takes it out of our head. Apps for our hand-held devices can help track sleep and even remind us of the optimal time for starting our bedtime routine. Some apps can sync with a wearable device such as a smartwatch to track our sleep and make further recommendations. Other apps can provide meditation routines or restful soundscapes. Of course, we want to resist that temptation to check our email. Perhaps for those who cannot resist, a white noise machine would be more appropriate.

Sleep Disorders

Researchers have identified five major sleep disorders. If the preceding recommendations have not been successful, seeking medical advice from your physician is the next best step. Some of these disorders are associated with serious medical conditions and require proper evaluation and treatment.

Sleep Apnea

Sleep apnea happens when your breathing is interrupted while sleeping, causing you to wake up suddenly. Surprisingly, this can occur hundreds of times in a single night. Not only will this reduce sleep quality, but it will also increase the risk of high blood pressure and heart attack.

REM Sleep Behavior Disorder (RBD)

More commonly known as "sleepwalking," REM sleep behavior disorder occurs when the natural body paralysis during REM sleep is absent. Although a rare condition, RBD can lead to physical injury. This condition usually requires medication and can be associated with an underlying neurological disease.

Insomnia

The differentiation between sleeping difficulties and true clinical insomnia can be a fine line. Insomnia is characterized by profound difficulty with falling asleep or staying asleep despite being exhausted. Causes include medications, stress, and mood disorders. After correcting for these causes, a standard treatment is cognitive behavioral therapy (CBT) involving practiced mindfulness and other behavioral techniques.

Restless Legs Syndrome (RLS)

Restless legs syndrome (RLS) is a condition that causes an uncontrollable urge to move your legs, typically occurring in the evening hours. Treatment can be very challenging. Ropinirole (Requip), rotigotine (NEUPRO), and pramipexole (Mirapex) are approved by the Food and Drug Administration for the treatment of moderate to severe RLS. Often, lifestyle changes such as warm baths, caffeine restriction, exercise, and foot wrapping are the best approach.

Narcolepsy

Narcolepsy occurs when excessive daytime sleepiness cannot be controlled, often causing one to suddenly fall asleep during the day. This can be a very dangerous condition, especially if it occurs while driving or operating machinery.

A Note About Sleeping Pills

Sleeping pills are seldom a safe long-term solution for sleep disorders and can often be addictive. Even nonprescription medications can impair cognitive function during the day. Data supporting effectiveness of antihistamines, CBD, and melatonin has been marginal.

So, What Happened to Demarcus?

Demarcus decided to make sleep a priority. He developed a ritual for the few hours leading up to a consistent bedtime. Within just a few weeks he felt more energized and even lost a few pounds of excess weight. Most importantly, his wife and children noticed a significant improvement in his mood when he came home from work. Perhaps this new routine will ensure enjoyable life experiences long beyond retirement.

Bottom Line

Sleep is a physiologic necessity. Lack of sleep will lead to a whole host of medical ailments, not to mention a diminished quality of life. Sleep deprivation is cumulative, and making up for this loss can take many nights. A couple of extra waking hours is seldom worth the price.

CHAPTER 21

Mental Health for Men

Jeff is a fifty-eight-year-old emergency room nurse who is feeling stressed and burned out at work. Jeff feels like he's always on edge, and he also is not sleeping well. He feels like his work is suffering, and he is worried about losing his job because of errors that he might make. Unfortunately, Jeff refuses to seek help for his problems because he thinks that seeing a mental health professional might be seen as a weakness. Instead, Jeff has been drinking two to three beers each night because it seems to relax him.

What Is Mental Health?

Let's start by defining mental health. The term describes a person's psychological, emotional, and social wellbeing. Mental health can affect the way we think and behave. It also affects the way that we interact with others. Many factors can contribute to our mental health, including the following:

- Genetics
- Family dynamics

- Childhood and life experiences
- Financial situation
- Physical health
- Lifestyle

Times are challenging and our mental health has suffered. Even before the COVID-19 pandemic, 20 percent of adults, a total of 50 million people in the US, reported having a mental illness.[1] When analyzing the statistics, it may appear that women suffer from mental illness more than men (25.8 percent vs. 15.8 percent[2]). However, the numbers don't tell the whole story. Women are more likely to seek treatment for mental health disorders. This is likely due to the stigma of mental illness and a concept known as "toxic masculinity." Men may not think that it's manly to admit to suffering from anxiety, depression, or other disorders. More than six million men suffer from depression alone in the US each year.[3] In addition, in 2020, almost four times as many men as women died by suicide.[4] Men are also more likely than women to abuse substances like alcohol and drugs.[5]

Mental Health Issues that Affect Men

Depression

Depression is a very common disorder in men, but it's not always recognized. Symptoms of depression can include persistent feelings of sadness or hopelessness, but it can also be characterized by:

- Feeling tired
- Having insomnia or sleeping too much

- Loss of interest in activities that were previously enjoyed
- Weight gain or loss
- Trouble concentrating
- Suicidal thoughts
- Becoming socially isolated from others
- Feeling angry or irritable
- Physical problems such as pain or digestive issues
- Sexual problems
- Engaging in risky behaviors or substance abuse

Men should seek help if they have ongoing symptoms. Having depression is not a weakness, and it will most likely get worse without treatment. Depression can improve with the right medication and/or psychotherapy.

Suicide

Almost forty-six thousand Americans died by suicide in 2020, making it the twelfth leading cause of death. This is sad news. Another sad statistic is that 79 percent of all suicide deaths in 2020 were by men.[6] Certain men are at higher risk, including the following:[7]

- Middle-aged men
- Men older than 65
- White men
- Men who live in rural areas
- Indigenous men
- Military veterans
- Incarcerated men

The Most Common Risk Factors for Suicide:[8]

- Depression or other mental illness
- Divorce or relationship problems
- Social Isolation
- Unemployment or financial problems
- Loss of a loved one
- Substance abuse
- Being bullied

The Warning Signs of Suicide:[9]

- Talking about death or dying
- Giving away possessions
- Making comments that life is not worth living
- Increased use of alcohol or drugs
- Decreased interest in socializing or participating in enjoyable activities
- Sudden change in mood

If you or someone you know is in crisis, call or text the National Suicide Hotline at 988 or chat online at 988lifeline.org.

Stress and Anxiety

Stress and anxiety are connected. Unfortunately, we all have stress in our lives. Stress is defined as "a person's emotional or physical response to the demands of life." Common causes of stress include financial problems, work issues, illness, relationships, and just daily life. Some stress is normal and can even be productive in helping you to meet deadlines or adapt to a new situation. Anxiety, on the other hand, is a feeling of worry, fear, or unease. It can occur as a result of a stressor, or it can happen without a cause. Anxiety is not a positive.

Symptoms of Stress and Anxiety

- Increased heart rate
- Feeling overwhelmed
- Trouble sleeping
- Being angry or irritable
- Having headaches or other pain
- Having digestive issues
- Experiencing a change in appetite
- Having trouble concentrating
- Feeling worried

Dealing with Stress and Anxiety

When mild stress or anxiety symptoms occur, the following strategies may help:

- Make sure that you are getting enough sleep (see Chapter 20).
- Make sure that you are getting enough regular exercise.
- Limit caffeine and alcohol.
- Find someone to talk to about the causes of your stress or anxiety.
- Practice mindfulness techniques like meditation or breathing exercises (see Chapter 17).
- Enjoy nature.
- Develop an attitude of gratitude.

You should seek the advice of a mental health professional if stress or anxiety is persistent or if it affects your daily life or ability to interact with others. Medications and therapy might be helpful to reduce symptoms.

Alcohol and Substance Abuse

Alcohol and drugs are related to mental health, and the relationship is bidirectional. People with mental health disorders are more likely to use these substances. In addition, those who drink or use illicit drugs are more likely to develop problems with anxiety, depression, or other threats to mental well-being.[10]

While alcohol and substance abuse are a problem for everyone, men are more likely to have addictions. According to the Centers for Disease Control and Prevention, close to 60 percent of men drink alcohol compared to 47 percent of women. Men are also more likely to binge drink, be hospitalized due to drinking, and die from an alcohol-related cause.[11] In addition, men are more likely to use marijuana, prescription opiods, and heroin.

Avoiding Alcohol and Substance Abuse

The best way to treat alcohol and substance abuse is to prevent it from happening. Here are some tips:

- Avoid triggers for drinking or using drugs. Know your limits when it comes to drinking.
- Don't take prescription medication that isn't prescribed by a health care professional for your personal use.
- When pain medication is prescribed for an injury or surgery, take only the lowest dose that eases pain, and take it for the shortest duration possible.

- Seek help for excess stress or mental disorders.
- Get support from family and friends.

Our takeaway for men's mental health:

- Mental health is suffering among men and women, and it's only gotten worse during the COVID-19 pandemic.
- Mental health is just as or even more important as physical health.
- Men should not be hesitant to reach out to seek help.
- A healthy lifestyle and self-care are important for men to reduce stress and improve overall mental health.

What Happened to Jeff?

Jeff's employer, a large hospital system, has developed a wellness plan for employees. They now offer a free gym membership, a meditation app, and free counseling services. Jeff's decided to utilize these services. He is seeing a counselor to talk about his work-related stress. Jeff has also started exercising three and four days every week and meditating for ten to twenty minutes daily. He feels much better, has stopped drinking beer at night, and as a result, is sleeping better. Jeff still has stress in his life, but he is much better able to manage it.

References

1. Mental Health America. "The State of Mental Health in America." Accessed July 20, 2022. https://mhanational.org/issues /state-mental-health-america.

2. National Institute of Mental Health. "Mental Illness." Last modified March 2023. Accessed August 1, 2022. https://www.nimh .nih.gov/health/statistics/mental-illness.

3. Chatmon, Benita N. "Males and Mental Health Stigma." *American Journal of Men's Health* 14, no. 4 (July-August 2020): https://doi.org/10.1177/1557988320949322.

4. American Foundation for Suicide Prevention. "Suicide Statistics." Accessed August 3, 2022. https://afsp.org/suicide-statistics/.

5. Chatmon, Benita N. "Males and Mental Health Stigma." *American Journal of Men's Health* 14, no. 4 (July-August 2020): https://doi.org/10.1177/1557988320949322.

6. American Foundation for Suicide Prevention. "Suicide Statistics." Accessed August 3, 2022. https://afsp.org/suicide-statistics/.

7. Whitley, Rob. *Men's Issues and Men's Mental Health: An Introductory Primer.* Cham, Switzerland: Springer Nature, 2021.

8. Kennard, Jerry. "Understanding Suicide Among Men." Very Well Mind (website), last modified October 3, 2022. Accessed August 3, 2022. https://www.verywellmind.com/men-and-suicide -2328492.

9. American Psychiatric Association. "Suicide Prevention." Accessed August 8, 2022. https://psychiatry.org/patients-families/suicide -prevention.

10. Mosel, Stacy. "Substance Abuse and Mental Health Treatment Programs." American Addiction Centers, last modified March 8, 2023. Accessed August 8, 2022. https://drugabuse.com/mental -health-drug-abuse/.

11. Center for Disease Control and Prevention. "Alcohol and Other Substance Use." Last reviewed July 25, 2022. https://www.cdc .gov/alcohol/fact-sheets/alcohol-and-other-substance-use.html.

CHAPTER 22

When Medical Advice
Is a Must

Reuben in a sixty-eight-year-old retired school administrator. Throughout his career, he has been a teacher, a coach, an assistant principal, and for the last fifteen years of his career, a school principal. He thoroughly enjoyed his work with his students and faculty. He retired at the age of sixty-five and enjoys good health. He and his wife, Helen, have been able to enjoy their passions in retirement, including more time for travel, reading, visiting grandchildren, and exercise.

Reuben started noticing some changes in his body at age sixty, and since then he has been on medication for his prostate, blood pressure, and cholesterol. At his last yearly physical, he commented to his physician, "You know, doc, after age sixty, it seems to be all tune-ups." His amused physician agreed with the truth in that observation. Considering some of the changes Reuben has started to experience, he is beginning to wonder what changes in his body warrant a trip to his physician, aside from his yearly exams. He has noticed an irregular heartbeat and he has lost some weight recently without changing his eating or exercise routines.

As we age, there will be changes in our bodies and in our health. Some of these changes are unexpected, and most are unwelcome, but such is the cost of maturity! We know our own bodies well, so when changes occur, a trip to a physician is in order. When a man experiences severe chest pain, a sudden fall, or unexpected bleeding, he knows that he must seek immediate medical care. However, a lot of aging changes are very slow and subtle in development and require our utmost awareness to realize there may be a health situation that deserves prompt attention.

Heart Disease and Cancer

The two leading causes of death in the United States usually remain the same every year. Heart disease is responsible for seven hundred thousand deaths, while cancer annually claims six hundred thousand lives.[1] Recently, COVID-19 has taken third place, followed by unintentional injuries, stroke, chronic respiratory diseases, Alzheimer's disease, diabetes, influenza and pneumonia, and kidney disease. Suicide is now the twelfth leading cause of death.

Since heart disease is the leading cause of death, you should know the signs and symptoms of early heart disease. Crushing chest pain, with the pain radiating into your left arm or into the jaw is a 9-1-1 situation that requires an immediate trip to the emergency center of the nearest hospital.

More subtle symptoms that should garner your attention are a decrease in exercise tolerance, mild chest pain, pain with exertion, shortness of breath, swelling in the feet or ankles, unexplained weight gain due to fluid retention, irregular pulse, and changes in your blood pressure. Any of these symptoms could be due to heart disease and warrant a trip to your physician. Many of these symptoms can be due to other causes, such as respiratory or kidney diseases. It will be up to your physician to provide you with a diagnosis and initiate a treatment to prevent heart disease. Your role is to see your physician.

Various cancers, of which there are around one hundred different types, can affect every organ in the body. The American Cancer Society (ACS) lists symptoms that *might* be warnings of your body having cancer. These symptoms include the following:

- Fatigue or extreme tiredness that doesn't get better with rest.
- Weight loss or gain of 10 pounds or more for no known reason
- Eating problems such as not feeling hungry, trouble swallowing, belly pain, or nausea and vomiting
- Swelling or lumps anywhere in the body
- Thickening or lump in the breast or other part of the body
- Pain, especially new or with no known reason, that doesn't go away or gets worse
- Skin changes such as a lump that bleeds or turns scaly, a new mole or a change in a mole, a sore that does not heal, or a yellowish color to the skin or eyes (jaundice)
- Cough or hoarseness that does not go away
- Unusual bleeding or bruising for no known reason
- Change in bowel habits, such as constipation or diarrhea, that doesn't go away or a change in how your stools look

- Bladder changes such as pain when urinating, blood in the urine, or frequency of urination
- Fever or night sweats
- Headaches
- Vision or hearing problems
- Mouth changes such as sores, bleeding, pain, or numbness[2]

The ACS emphasizes that these signs or symptoms *could* be due to cancer, but often when these symptoms occur, it is a cause other than cancer. Nonetheless, if an individual is experiencing these types of symptoms, especially over weeks or months, a consultation with your physician is in order. The list above is not all-inclusive, and there are others than can occur, but the important point is to be aware of *any* changes and report them to your doctor.

We've looked at some of the warning symptoms of the "big two," heart disease and cancer. There are other systems in the body with warning signs that you must be aware of and heed.

The Skeletal System—Your Bones and Joints

Your bones and joints and our skeletal system are used every day with every move you make. As you age, you can expect some wear and tear and mild aches and pains. There are more than two hundred bones in the human body and most are separated by a joint, so there are many places among those bones and joints for things to go wrong.

Orthopedic surgeons are very busy repairing and replacing hips, knees, and other joints in tens of thousands of patients every year. We suggest you seek help from an orthopedic surgeon regarding skeletal issues such as a fall with a break, a sudden change in mobility, or a bone out of place. More subtle, slowly developing skeletal changes may not seem to be more than minor annoyances, but many of the aging changes we experience are likely to become progressive and more difficult to treat or cure if you don't seek help early on.

As part of aging, almost everyone will develop *osteoarthritis,* due to long-term wear and tear of our joints. Osteoarthritis is different from *rheumatoid arthritis,* which is an autoimmune disease. (More on that later.) Be aware of the subtle changes in mobility, strength, gait, joint tenderness, redness, or swelling and contact a physician, either primary care or an orthopedic surgeon for assessment of these issues. Early diagnosis and early intervention are generally better than letting a bone or joint problem continue to worsen. In the past, some joint conditions such as rheumatoid arthritis were destined to progress to severe joint deformity and accompanying disability. Rheumatoid arthritis and psoriatic arthritis are autoimmune diseases wherein our immune cells attack and damage our own tissues destroying healthy joints. These conditions were previously untreatable, but today we have successful treatments of these autoimmune diseases. We recommend that you be aware of even subtle changes and proceed to early evaluation and treatment.

The Endocrine System—Our Glands

Our endocrine system comprises the organs that secrete hormones. Hormones are just one of many substances made by your body's glands. They circulate in the bloodstream and control the actions of certain cells or organs. Hormones are secreted by the pituitary gland, the adrenal glands, the thyroid, parathyroids, and the pancreas, which make up the bulk of our endocrine system. (The testes which produce testosterone and are discussed extensively in Chapter 7).

Two endocrine glands that get the most medical attention and cause the greatest number of medical problems for large numbers of the population are the thyroid gland and the pancreas.

The thyroid gland is located in the front of your neck; it produces thyroid hormone. Proper levels of this hormone are under the influence of the pituitary gland. When your thyroid produces too much or too little of its hormone you will experience a wide range of symptoms, both minor and major.

Your pancreas is located in your abdomen and is connected to your gastrointestinal or digestive tract. When it is functioning properly it produces the right amount of insulin to maintain your blood sugar (glucose) levels in a proper range; in addition, the pancreas produces various chemicals that facilitate your food digestion.

Although these two hormone-producing glands play different roles in maintaining your physiology, malfunctions of both glands can produce similar symptoms that should get your attention, including a visit with your physician.

A hyperactive thyroid, hyperthyroidism, will lead to unexplained weight loss and an irregular heart rhythm. An underactive pancreas, failing to produce adequate insulin, causes type 2 diabetes mellitus, which will also lead to unexplained weight loss.

Excessive urination and thirst are also some of the early warning signs of diabetes. There are other causes of weight changes, and changes in heart rhythm and rate, but any of the symptoms listed above, especially if persistent, should motivate a visit to your primary care doctor or an endocrinologist for evaluation. Our take-home message is that any *unexplained persistent changes in your body deserve medical attention.*

The Nervous System

Our nervous system is a marvelous thing, as all our bodily activities depend upon the proper function of our nerves. The nervous system consists of these parts: the brain; the twelve cranial nerves that affect most aspects of the function of your head, face, and neck; your spinal cord that runs from your brain to your tailbone; and the nerves that leave your spinal cord and affect the function of all your internal organs. Any miscommunication between the nerves in the system is going to affect some bodily functions.

The most well-publicized adverse event in the nervous system is a stroke. This occurs when there is either bleeding within the brain from a broken blood vessel or when the blood supply to the brain is impaired by a blocked blood vessel. Previous evidence of a stroke is sometimes seen with a brain imaging test and there are findings of a previous stroke that had no clinical signs or symptoms. More common, however, is the dramatic stroke that causes symptoms which you cannot ignore. There are five classic warning signs:

- Sudden numbness or weakness in any part of the body;
- Sudden confusion or trouble speaking or understanding speech;
- Sudden vision problems;

- Sudden difficulty with walking;
- Drooping of the face muscles; an uneven smile.

Any of these symptoms could be due to a stroke, and each is considered a medical emergency that requires a 9-1-1 call. If a stroke is diagnosed and treated early, the damage can be stopped or even reversed.

Many neurological diseases only cause subtle symptoms. And most neurological diseases, such as multiple sclerosis (MS), Parkinson's Disease, and Amyotrophic Lateral Sclerosis (ALS or Lou Gehrig's Disease*) are progressive. Unexplained changes in strength, gait, cognition, and nerve pain such as sciatica with severe pain tracing down the back of your leg, warrant a visit with your physician. You know your own body, and being alert to subtle changes is the key to early intervention, diagnosis, and treatment of conditions of the nervous system.

Vision and Hearing

As we age, we will suffer some loss of hearing and vision. Look around and notice how many elderly men and women are wearing hearing aids. Cataracts caused by fogging or clouding of the lens in the eye are almost inevitable if we live long enough. A cataract is easily removed by an ophthalmologist and is replaced with a plastic lens as an outpatient procedure that restores vision. These are common aging events. We

---

* This is the only disease named for a person who suffered from the disease—Lou Gehrig. Other names of diseases such as Parkinson's, Graves, Huntington's, and Peyronie's are names of individuals who either first described the disease or were very instrumental in research regarding the disease.

recommend that if you are over fifty years of age, that you see an eye specialist on a yearly basis. An ophthalmologist will screen for some vision-robbing conditions such as macular degeneration and glaucoma. Both macular degeneration and glaucoma in its early stages may be asymptomatic and if not detected can lead to complete loss of vision. This is especially true if you are diabetic. We also recommend that you have your hearing system checked periodically by an otolaryngologist to detect subtle problems that could be managed by early intervention.

Certain issues with the eyes and ears demand *prompt* attention. As with so many other systems in the body, it is the acute and unexpected change that demands an immediate medical attention such as sudden changes in vision or sudden pain in the eye. The same can be said for the hearing system, which includes sudden unexplained change in hearing acuity, or severe ear pain. These sudden changes do not always indicate something serious, but needs prompt attention. Our take-home message it to be aware of your body and alert to subtle changes that may require medical evaluation, but *acute and dramatic changes may demand immediate attention*.

The Gastrointestinal System

Our system of digestion begins working the moment we put food into our mouths. Saliva starts the process, and the process continues for another twenty feet through the esophagus, stomach, duodenum, small intestine, large intestine (colon), and finally, what we don't utilize as nutrients, exits the rectum. All along the way, our bodies are absorbing the nutrients we need, and expelling what we don't need. Two important organs, the liver and the pancreas, attached to the intestinal tract just beyond the stomach, facilitate digestion. Further along, at the junction of the small intestine with the colon, is the appendix, which has no useful purpose, but can develop appendicitis and an emergency medical situation requiring its surgical removal.

Probably all of us suffer some minor gastrointestinal upsets such as indigestion, a stomach ache, diarrhea, constipation, nausea, and even vomiting from food that didn't agree with us. If these symptoms are short-lived, no physician intervention may be needed. However, if any of these symptoms are persistent or severe, schedule a visit with your physician.

Some signs within the gastrointestinal system that could portend a potentially serious condition include changes in the stool. If the stool is white or black or red, these colors are not normal and could indicate disease of the liver, stomach, or colon. Any change in the color of the stool requires prompt physician attention. If you were to notice the whites of your eyes, the conjunctiva, appearing yellow this may indicate liver disease, gallbladder disease, or even pancreatic cancer. These are serious symptoms. Acute pain in the abdomen that does not resolve needs attention. Diverticulitis, an inflammation of the colon, can cause severe damage with colon rupture and usually presents as significant pain and tenderness in the lower left abdomen. Appendicitis usually presents the same way, but on the right side. Pain, distention, and vomiting can be signs of intestinal blockage. All such symptoms usually warrant an urgent trip to the hospital. These are not symptoms you want to "tough it out."

As you have seen in this chapter, the emphasis is on changes in your normal gastrointestinal function. You know your body, and you know when things aren't right. Always be alert and vigilant. One extra trip to the physician is better than one too few.

Dermatology and Your Skin

Your skin is your largest organ. It protects you from the outer insults to your body: heat, cold, cuts, infections, sunlight, bumps, and bruises. Skin has the capacity to regenerate itself. Old skin cells die and new ones replace them.

Skin ailments such as rashes, bumps, bruises, and acne are self-limiting and of minimal concern. However, skin cancer is a condition that does require attention. If you have a new spot (lesion) that does not clear in a short time, a trip to your primary care physician or your dermatologist is in order. An annual skin check by a dermatologist is appropriate for men beginning at ages twenty to thirty.

Skin cancers are not often obvious, but a lesion that is different in color from your normal skin, becomes larger, and exhibits an irregular border is a warning that a physician visit is in order to be certain that you don't have a treatable skin cancer. It may be nothing, but you want to be sure. Often the dermatologist cannot be sure without a biopsy of the lesion. Melanoma is particularly dangerous, even fatal, so do not ignore questionable skin lesions.

Our skin is going to change as we age. That nice soft, smooth, blemish-free skin of our youth changes into the skin of aging. We expect that unfortunate fact of life, but changes that don't seem to be a normal part of aging deserve our close attention and perhaps a physician's visit. Knowing our bodies and being keen on changes, and not ignoring them, is the key to successful aging of your skin.

Back to Our Friend, Reuben

As you recall, Reuben, a retired school administrator, enjoyed good health, but at age sixty-eight had noticed some unwanted bodily changes. He felt his heartbeat from time to time seemed fast and irregular. There did not seem to be any obvious explanation, and it did not seem to be related to his exercise program. Over a period of several months, he had lost fifteen pounds without any explanation. He didn't feel unwell, but this was not the normal Reuben. His physician agreed that these changes were unexpected and needed evaluation, explanation, and possibly treatment. It did not take his physician long to do some tests and receive answers for Reuben. With changes in his heart rate and unexplained weight loss, the possibility of overactive thyroid and diabetes were high on the list of possibilities.

An EKG revealed a heart irregularity, and blood tests confirmed hyperthyroidism. Additional testing revealed no evidence of diabetes. Reuben is now on thyroid treatment and his cardiologist is working on his atrial fibrillation, a common heart irregularity, especially in the face of excess thyroid hormone. Fortunately, Reuben was paying attention to his body and the changes he was experiencing and sought the help of his physician before things progressed and became worse and more challenging to manage.

Bottom Line

This chapter is not designed to be an encyclopedic discussion of all the reasons a man should seek medical care, especially urgent care, but it is designed to acquaint you with many symptoms and ailments which at first might seem minor or worth ignoring, but in fact, should prompt a visit to a physician or even an emergency center visit. As has been stated several times throughout this chapter, and still deserving of one more mention: be alert and aware of changes in your body. We expect many unwanted changes as we age, but it is important to notice changes that seem unexpected and beyond the usual process of aging. If there is a question in your own mind as to whether you need to visit a physician, better to err on the side of one unnecessary visit than to avoid the one visit that might turn up something very important to your health, life, and longevity.

References

1. Ahmad, Farida B., and Robert N. Anderson. "The Leading Causes of Death for 2020." *JAMA* 325, no. 18 (May 2021): 1829–30. https://doi.org/10.1001/jama.2021.5469.
2. American Cancer Society. "Signs and Symptoms of Cancer." Last modified November 6, 2020. https://www.cancer.org/treatment/understanding-your-diagnosis/signs-and-symptoms-of-cancer.html.

Conclusion

It is a fact that women tend to live longer than men. The average American man will live to age seventy-six, while the average woman in America will live to age eighty-one. We have recognized this age disparity and want to provide men with ways to narrow that gap. Why are women living longer than men? We have attempted to answer that question by writing this book.

There are a few theories—some to do with biology and some with behavior. Men are more likely to smoke, drink excessively, drive too fast, and be overweight. Our practices have provided care for thousands of men, and we have noted that men are less likely to seek medical help early. They are more likely to be non-adherent to treatment if diagnosed with a disease. In addition, men are more likely to take life-threatening risks. We believe that the gender age gap can be narrowed if men lead healthy lifestyles and avail themselves of the healthcare profession.

Women develop relationships with physicians while teenagers and even more so when they become pregnant. Conversely, men often don't start a relationship with physicians until their late forties or fifties. During that time, they may have problems with their blood pressure, weight, and other diseases that remain silent for many years. Women start screening for breast and cervical cancer at an early age while, at the same time, men avoid screening entirely or wait until they start having symptoms.

Men frequently think because they feel fine, there's nothing to worry about. The "If it ain't broke, don't fix it!" theory may work for cars, but not always for men. Men are often not prevention conscious. Still, men put off going to the doctor, even if they suspect something might be wrong. Most men often fear the doctor. Men are metaphorically like ostriches, with their heads in the sand, not oriented to seek help but believing that changes in bodily function will temporarily subside given time.

We have tried to emphasize that nearly 100 percent of men of any age will sometimes have unexpected medical issues and physical challenges. As you age, many changes will occur in your body and ultimately, your health. Many of these changes are unexpected and unwanted. These changes are frequently a complete surprise. We cannot count the number of times we have heard our patients say, "But Doctor, I've never had this before."

This is life as time moves on. In this book we have tried to prepare you for these life/body changes so they can be understood as they occur, and so that you will have the tools to ward off *some* of these changes through good life habits.

As you age, and your body changes, we want you to know that you should not suffer in silence; instead, make every effort to be proactive in your health. We have written this book because many men are uncomfortable discussing these issues with their friends, partners, and even physicians. The truth is that the unique male organs between your belly button and your knees can be a source of many problems we have discussed in this book. But troubles can be lurking down there, and elsewhere, without being obvious.

This book represents a compilation of stories from men like you who have experienced similar challenges and uncertainties down there, as well as in other organ systems throughout their bodies. We want you to recognize that information and appreciate that help is available. Your road to recovery and good health begins with understanding your problem and the underlying symptoms, pain, or discomfort you are experiencing. The next step is to speak to your doctor, who will craft a plan of action or refer you to a male health specialist or other specialists who can help you.

The path to improvement will require you to execute your personal plan of action. Remember, to know and not to do is tantamount to not knowing. You must partner with your doctor and follow the course of action laid out for you. Only then will you gain control of the issues with your body that are preventing your return to full health.

We want this book to serve as a guide to help make you a happier and healthier man. Hopefully, you will read this book—or at least the chapters that are important to you—before your appointment with your physician. After reading, it is our hope you can better understand your physician's advice, and that you will have the requisite knowledge to help your physician reach an accurate diagnosis and start you on your road to recovery.

We want to empower you to develop a support system with which you can discuss what is going on throughout your body, whether it's your physician, partner, friends, or relatives. It's important to feel comfortable and confident sharing your travails and triumphs, seeking advice from those you trust, and sharing your experiences with those who are still striving to understand what is going on down there—and elsewhere—throughout our bodies.

Finally, thank you for reading our book. We would like to hear from you. If you have a success story and would be willing, please share it with us so we can help others learn from your experience.

doctorwhiz@gmail.com for Dr. Neil Baum
houseofmiller@me.com for Dr. Scott Miller
mindi.miller@hotmail.com for Dr. Mindi Miller
mobleyresearch@gmail.com for Dr. David Mobley

About the Authors

Neil H. Baum, MD, is Professor of Clinical Urology at Tulane Medical School in New Orleans, Louisiana.

Dr. Baum has written more than 250 peer-reviewed articles and seven books on men and women's health care issues. *Ecnetopmi-Impotence, It's Reversible* and *What's Going on Down There: The Complete Guide to Women's Pelvic Health*, for women.

Dr. Baum was the columnist for *American Medical News* for over twenty-five years. For more than twenty years, he also wrote the popular column, "The Bottom Line," for *Urology Times*. Dr. Baum also focuses on patient education and has written extensively on enhancing communication with patients and their families.

Dr. Baum has made an effort to simplify the communication with patients who have concerns about prostate cancer screening, erectile dysfunction, and urinary incontinence.

Dr. Baum is married to Linda, has three children, Alisa, Lauren, and Craig, and six grandchildren.

Scott D. Miller, MD, MBA, is a men's health expert who believes patients should be informed and involved in their care. He has dedicated his career to educating the community and his colleagues through original writings and regular television and radio appearances. As a board-certified urologist for over twenty years, Dr. Miller has one of the largest, most diverse experiences in laparoscopic and robotic urology in the Southeast. He is an advocate for minimally-invasive techniques so that patients can quickly return to their normal routine.

Dr. Miller practices urology at Wellstar North Fulton Medical Center and is the medical director of robotic surgery and medical director of urology at Wellstar Health System. He is the founder and president of ProstAware, a non-profit prostate cancer awareness organization. He is also an accomplished musician and composer who has written a variety of songs, including one about the emotional impact of a cancer diagnosis. He lives in Atlanta, Georgia, with his wife Mindi, a co-author of this book.

Mindi S. Miller, PharmD, is an Associate Clinical Professor at the University of Georgia College of Pharmacy. She is a native of Atlanta, Georgia, and graduated with a BS in Pharmacy from UGA and a Doctor of Pharmacy degree from the University of Kentucky. Dr. Miller is a Board-Certified Pharmacotherapy Specialist (BCPS). She began her career as a staff pharmacist and later served as clinical staff pharmacist in the area of critical care at the University of Kentucky Medical Center in Lexington, Kentucky. Dr. Miller also worked at Emory Healthcare for twenty-five years, where she practiced clinical pharmacy in the areas of cardiology, pharmacokinetics, and nutrition support.

Dr. Miller has been involved in pharmacy education for over thirty years and has enjoyed working with both residents and students. Dr. Miller specializes in the areas of men's health and overall wellness and is the author of several publications on those topics. She also teaches courses in men's health and wellness for UGA College of Pharmacy. She recently completed a certificate program in wellness sponsored by the American Society of Health-System Pharmacists (ASHP). She is married to Dr. Scott Miller.

David F. Mobley, MD, FACS is Associate Professor, Clinical Urology with Weill Cornell Medicine and Houston Methodist Hospital. Throughout his career, he has been extensively involved in patient care as well as medical research, having been a Principal Investigator in more than one hundred FDA-approved clinical trials of drugs and medical devices. His work has been extensively published in medical journals. He has also authored chapters in medical textbooks and has published two books on erectile dysfunction and a recent book on prostate cancer. As a result of his surgical experience, research, and clinical investigations, Dr. Mobley has been devoted to teaching urologists and lecturing at medical seminars both domestically and internationally for over four decades.

Dr. Mobley is one of a small number of urologists offering HIFU treatment for localized prostate cancer. For over fifteen years, he has been extensively involved in this minimally-invasive and non-surgical heat treatment for prostate cancer.

Dr. Mobley continues practicing urology in Houston, Texas, where he has practiced for over forty-five years.

Index